GENETIC SCIENCE AND NEW DIGITAL TECHNOLOGIES

Science and Technology Studies and Health Praxis

Edited by
Tina Sikka

BRISTOL
UNIVERSITY
PRESS

First published in Great Britain in 2023 by

Bristol University Press
University of Bristol
1-9 Old Park Hill
Bristol
BS2 8BB
UK
t: +44 (0)117 374 6645
e: bup-info@bristol.ac.uk

Details of international sales and distribution partners are available at bristoluniversitypress.co.uk

British Library Cataloguing in Publication Data
A catalogue record for this book is available from the British Library

ISBN 978-1-5292-2331-6 hardcover
ISBN 978-1-5292-2332-3 ePub
ISBN 978-1-5292-2333-0 ePdf

Cover design: Nicky Borowiec
Front cover image: Susceptible by Anna Dumitriu and Alex May
Bristol University Press uses environmentally responsible print partners.
Printed and bound in the UK by CPI Group (UK) Ltd, Croydon, CR0 4YY

FSC
www.fsc.org
MIX
Paper | Supporting
responsible forestry
FSC® C013604

Contents

Notes on Contributors

Filomena Berardi is a Marie Curie PhD Social Psychologist with a focus on inter group relations currently working as a researcher at Ires Piemonte (the Institute of Economic and Social Research) for the evaluation of EU projects supported within the EU Cohesion Policy. In particular, she focuses on the effects of innovation projects and on dynamics of collaborative innovation within Regional development. She also supports capacity building initiatives addressed to public servants with the goal of implementing local authority's capacity to attract European resources and funding.

Kevin Cummings is Professor in the Department of Communication Studies and Theatre Arts at Mercer University. His research explores intersections between rhetoric and media, including scholarship on artificial intelligence, invasive species, and Twitter.

Miquel Domènech is Senior Lecturer in Social Psychology at the Universitat Autònoma de Barcelona. His research interests cohere broadly in the field of science and technology studies (STS), with a special focus on the relationship between care and technology and on citizen participation in technoscientific issues. He is the coordinator of the Barcelona Science and Technology Studies Group (STS-b).

Anamika Gulati is Assistant Professor in the Centre for Studies in Science Policy at Jawaharlal Nehru University. Her research interests are in medical affairs and scientific communication, clinical trials, health outcome research, health policies in the field of STS, modern science and traditional knowledge, and biomedical waste management, among others.

Daphne Oluwaseun Martschenko is Assistant Professor at the Stanford Medicine Center for Biomedical Ethics. Her work focuses on the ethical, legal, and social implications of human genetic/genomic research. Through her scholarship she advocates for and facilitates socially responsible research, communication, and community engagement in social and behavioural genomics.

Elizabeth Mills is Senior Lecturer in Social Anthropology and International Development in the School of Global Studies at the University of Sussex. Using visual anthropological approaches – including film, photography, body mapping and digital mapping – her ethnographic research has explored women's embodied accounts of HIV and AIDS biomedicines in South Africa and Brazil.

John Rief is Associate Professor in the Department of Communication Studies at Metropolitan State University of Denver. His areas of expertise include public speaking, rhetoric, argumentation theory and practice, deliberation, competitive and civic debate in the US and best practices of international and intercultural debate.

Celia Roberts is Professor in the School of Sociology in the College of Arts and Social Sciences at Australian National University. She works in the area of Feminist Technoscience Studies, with particular focus on reproduction, sexuality, sex/gender, embodiment and health. She is co-author of a forthcoming book, *Reproduction, Kin and Climate Crisis: Making Bushfire Babies* (Bristol University Press).

Kazuhiko Shibuya is Professor in the Department of Global Business at Globiz Professional University, Japan. His research methodologies are mainly social system design, software development, agent-based social simulations, social survey and statistical analysis.

Tina Sikka is Reader in Technoscience and Intersectional Justice in the School of Arts and Culture at Newcastle University. Her current research includes the critical and intersectional study of science, applied to climate change, bodies and health, as well as research on consent, sexuality and restorative justice.

Núria Vallès-Peris is a postdoctoral researcher, thanks to the Margarita Salas grant at the Intelligent Data Science and Artificial Intelligence Research Center of the Universitat Politècnica de Catalunya, and is also a member of the Barcelona Science and Technology Studies Group (STS-b) and the Institut d'Història de la Ciència of the Universitat Autònoma de Barcelona. Her research focuses on the study of the ethical, political and social controversies surrounding robotics and artificial intelligence, especially in the fields of care, health and public policy.

Giorgio Vernoni is a researcher in labour economics at IRES Piemonte (the Institute of Economic and Social Research of the Piedmont Region) and is a partner of the LABORatorio Riccardo Revelli – Centre for Employment

Studies. His main research interests focus on the relationship between technology and work, regional labour systems, skills needs analysis, agency work, labour market policies and employment services, policy evaluation, and non-traditional and administrative data sources.

Joann Wilkinson is an academic advisor in the School of Social Sciences at the University of Manchester. Her research interests include gender, health, self-tracking and data practices.

Introduction

This collection began as a personal undertaking aimed at bringing together a diverse set of scholars whose work on health and technology converged on a shared objective: namely, to investigate a slice of our contemporary health-centric sociotechnical world using intersectional Science and Technology Studies (STS) methods in innovative ways. The first spark for this project came from my work on another book in which I examined the material-discursive construction of health and STS, titled *Health Apps, Genetic Diets and Superfoods: When Biopolitics Meets Neoliberalism*. In the course of writing that book, I discovered a flourishing ecosystem of new and exciting research examining what being in 'good health' means for individuals and communities and how this intersects with more contemporary notions of well-being in light of technological change, increasing medicalization and biomedicalization, and controversies around care, access, sovereignty and justice (Henry, Oliver and Winters, 2019; Hatch, 2020; Kolopenuk, 2020; Flore, 2021; Lupton and Willis, 2021).

Also of note was how health, under the conditions of a pandemic, had taken on a new valence – one that thematized the depths of our relations with other humans, non-humans and artefacts (inclusive of technologies). It also marked, as Moya Bailey and Whitney Peoples contend, a reconceptualization of health as 'both a desired state of being and a social construct necessary of interrogation because race, gender, able-bodiedness, and other aspects of cultural production profoundly shape our notions of what is healthy' (Baily and Peoples, 2017, p 3). While not all the chapters in this volume take on health as discourse and embodied state of being, each, in its own way, weaves together compelling narratives and performs agential cuts into our mutating techno-scientific and health assemblages in novel and exciting ways (Barad, 2007).

In light of these revelations, in drafting a proposal and sending out a call for submissions, I made it clear that my objective in inviting contributors and soliciting chapters was to use the book to engage in conversations with a diverse set of scholars whose commitment to heterogeneous scholarship was not only reflected in the thinkers they draw on, but also their methods, choice of case studies, writing styles and intentions.

This commitment to disruption requires that we 'trouble' traditional research and engage with the larger world as a 'spacetime mattering' that is 'complicated, compromised ... impossible to conceptualize' (Shotwell, 2016, p 10; also see Bailey and Peoples, 2017) and compels 'us to make continually contingent and unsettled decisions about how to be in relation to the world, with no predetermined answer' (Shotwell, 2016, p 196; also see Barad, 2007). In the remainder of this introduction, I provide a basic overview of STS and its significance, followed by a summary of the chapters that make up this book.

The importance of STS

By way of background, contemporary STS as a method and field of research has evolved from its post-Kuhnian (1962) roots to focus on the larger ecosystem of scientific practices and technological artefacts. One of its most enticing attributes is the sheer number of sociotechnical spaces on and around which STS-focused approaches can be applied. Some of the most compelling work in STS I have come across in the past few years has been in unexpected areas such as a 2012 study of podcasting devices in Zimbabwe examined through the lens of design and capabilities (Oosterlaken, Grimshaw and Janssen, 2012) and a 2020 article assessing the place and role of race in studies of the microbiome (Benezra, 2020).

Vis-à-vis epistemology, STS focuses on exploring knowledge-making practices, whether that is in the lab (Latour and Woolgar, 1979); knowledge-creating communities (Jasanoff, 2004) with respect to social effects (Hackett et al, 2008); the dynamics between science and technology (Bijker and Law, 1992); scientific controversies (Pint and Leuenberger, 2006); or public understandings of science and technology (Jasanoff and Kim, 2013). While there is no one method of practising STS per se – something which tends to cause consternation among students – STS scholars can and do draw on a suite of largely qualitative methodologies including auto-ethnographies (Rudenko, 2016), actor-network theory (Latour, 2005), ethnomethodology (Sormani, Alač, Bovet and Greiffenhagen 2017), network analysis (Venturini, Munk and Jacomy, 2019), new materialism (Lemke, 2015), feminist STS (Bauchspies and de La Bellacasa, 2009), post-colonial STS (Lyons et al, 2017), Black STS (Benjamin, 2016), queer STS (Molldrem and Thakor, 2017) and disability STS (Blume, Gailis and Pineda, 2014). These approaches reflect some of the more recent ways in which STS has developed and in which power, inequality, identity and affectivity have come to take on an important valence. This research continues to offer exciting ways to think through and with science and technology as discursive-material practices that are constructed and circulate within particular contexts – ones that are raced, classed, sexed and gendered.

In addition to the study of health, methods and epistemology, this collection also aims to make a contribution to the 'science *and* technology of it all' by thinking with and through the most recent and novel processes, ideas, institutions, pedagogies, values, assumptions, principles and artefacts aimed at reconstructing what it means to be human – whether it is through extensions *of self* or transformations *to self*. Of central concern is how technologies might be shaped so as to (re)configure knowledge, (re)configure connectivity and (re)configure control (Henwood and Marent, 2019).

The unifying thread that ties these chapters together is a commitment to understanding whether and how scientific principles and health-oriented technologies might be gathered in ways that contribute to a 'good', socially rich, healthful and just life. Sometimes this might involve the considered adoption of new technologies, or, alternatively, their total deconstruction alongside plans to rebuild.

Chapter outline

Chapter 1, written by Daphne Martshenko, engages in a study of genomics through the lens of discriminate biopower. Martshenko takes the case of education, a critical wellness indicator, as a locus through which emergent genomic technoscience is being developed, ostensibly, to enhance educational outcomes. However, as Martshenko makes clear, these technologies are more likely to maintain the status quo or, in the worst-case scenario, exacerbate 'educational inequalities' which cleave along race, gender and class lines in ways that are 'enhanced and regularized through "molecularized" sciences'. She calls, instead, for the cultivation of a new imaginary based on equitable resource allocation and social justice rather than one rooted in biological essentialism and genetic determinism.

In Chapter 2, my own chapter, I draw on feminist STS and feminist new materialism (FNM) to construct what I call an 'auto-ethnographic and rhizomatic socio-material feminist approach to science and technology' to examine 'immunity boosting supplements' whose market share grew exponentially as a result of the COVID-19 pandemic. I apply a generative approach along with FNM to study Purearth, a UK health and wellness company. I assess the discursive-material entanglements between body normativity, immunity and neoliberalism using agential cuts to explore how dominant health norms are reinscribed and perpetuated. I also, drawing on auto-ethnographic STS, enact my own 'experiment' by consuming their supplements for 10 days during which I co-produce insights and findings around cultural appropriation, gender, race and their intersections. As this is one of two chapters that engages in auto-ethnographic work, I think it is important to draw attention to this method as forming an important part of contemporary STS scholarship (Adams et al, 2018; Vigren and Bergoth,

2021). I conclude the chapter by calling for a fundamental reconceptualization of health as it relates to superfluous supplements by cultivating connections and commitments to movements like Black veganism, fat justice and health sovereignty, which, I contend, hold more promise.

Elizabeth Mills, in Chapter 3, assembles a multifaceted FNM and auto-ethnographic analysis of anti-retroviral (ARV) use in South Africa. She draws on the lived experiences of women engaged in community activism and receiving ARVs whose rich accounts of these potentially life-saving drugs are demonstrably more complex than one might think. The ARVs themselves are portrayed as complicated social actors whose effects are viewed through a post-colonial lens and in line with both sociopolitical and affective forces and flows. Mills demonstrates how contemporary biomedical models of health need to be understood as emerging out of a racialized and colonial context, as operating in 'a gendered and unequal social order', and as intra-connected with embodied experiences of illness that 'push beyond the binary of either being healthy (and embodying technologies "successfully") or being ill (with bodies that are "failing")'.

Kazuhiko Shibuya's Chapter 4 engages in a wide-ranging discussion of artificiality as it relates to cutting-edge technological innovations like AI, genetic engineering and nuclear energy. The subjects of existential anxiety, progress, morality, threat and risk are taken up in order to prise apart our contemporary technological lifeworlds and think through how new technologies function, as well as how they might do so differently. Dignity, controllability and legal authority, Shibuya argues, are of paramount importance, as are equity and privacy. He closes his chapter with a comprehensive call for inclusive design, open engineering and sustainable futures.

Kevin Cummings and John Rief's Chapter 5 acts as a compelling rejoinder to Shibuya's piece by examining the ways in which a humanized and humanizing AI might function to co-produce a more ethical health ecology. Building on the communicative capacity and affordances of AI, and pushing back against the tendency towards technophilic quantification and medicalization, they argue that AI can be constructed and deployed in ways that are rehumanizing and patient centred – particularly when seen as part of a larger medical 'pit crew' working together to enact a deep form of medical relationality. Drawing on scholarship from rhetorical studies and using a deliberative approach, Cummings and Rief make the case that 'AI should be a full-fledged member of the healthcare team with the goal of humanizing not only healthcare practitioners and their patents but also the technology'. This fits nicely with STS approaches in which technologies are given agency as non-human actants by scholars like John Law (2004), Michel Callon (Callon and Law, 1997) and Bruno Latour (2000).

Chapter 6, by Anamika Gulati, draws on the Indian context in order to explore how gender biases operate in the practice and communication of science and technology. The chapter opens with a discussion of digital health technologies before moving to the case of women in STEM. Vis-à-vis health, Gulati examines the subjects of privacy, inequality and public benefit in order to draw out how health data can be used with integrity. She then uses feminist STS to make the case for a capacious and socially just approach to science communication that includes open access, state support, targeted resources and improved scientific literacy. The place of women in these debates guides the chapter, which concludes with a call for a 'better understanding of the proliferating paths of research and the increasing effects of interventions pulled by feminist concerns'.

In Chapter 7, by Filomena Berardi and Giorgio Vernoni, the authors examine the myriad obstacles to the adoption of medical technologies, using radiological AI as a case study. Drawing on the sociotechnical as an analytic frame, as well as semi-structured interviews, Berardi and Vernoni explore key contradictions as well as the role played by fear and lack of trust in preventing technological uptake. The roles of automation, AI, economic systems and social norms are then used to produce a contextual account of technology adoption that is multi-perspectival and provides a robust explanation of 'automation in medical imaging, paying particular attention to the relationships between doctors and between doctors and patients'.

Miquel Domènech and Núria Vallès-Peris' Chapter 8 on 'robots for care' addresses arguments for and against the use of robots in medical contexts and particularly as it relates to care work. Arguments supporting their deployment include the so-called care crisis, the demand for efficiencies, new potentialities (for autistic children for example) and market potential. They make a case in favour of heeding ethical concerns and use STS to round out the critical part of their analysis, including the capacity for robots to transform social relations, their tendency to obscure or 'black box' their operations and their predisposition to concretize society's assumptions and values. Domènech and Vallès-Peris close their chapter on the subject of design, an important element of contemporary STS scholarship, which they argue requires the participation of social scientists *and* users from inception. They emphasize strongly that what is needed is for decision making around care robots to guarantee that chosen values (care, dignity, independence) are strengthened, social relationships cultivated and caregiving improved.

Finally, in Chapter 9, Joann Wilkinson and Celia Roberts deploy scholarship in feminist STS to explore the timely question of whether ovulation biosensors are feminist technologies. Drawing on ethnographic interviews and applying a FemTech lens, Wilkinson and Roberts unpack how modern ovulation technologies are constituted by the conflicting discourses of agency and empowerment on the one hand, and neoliberal responsibilization on the other. Also discussed are concerns around privacy, data protection and expertise – which are themes reflected in each of the chapters in this collection. The rich narrative accounts provided by the women interviewed speak to the sophisticated and complex nature of the intra-actions that occur between their embodied sensory experience and ovulation biosensors. These interfacing events allowed some of the women to forge a feminist relation with their devices such that they were able to 'push against the normative boundaries enacted by biosensing technologies to learn to see all kinds of bodies in different ways' and, in doing so, co-produce their own 'expert' knowledge. And yet, like other health technologies, these devices have been built in line with capitalist agendas and individualizing objectives about which Wilkinson and Roberts express concern.

Taken together, these chapters provide a robust, insightful and rigorous examination of health technologies, whether we are talking about medical imaging, AI, ARVs, biosensors or supplements, using novel applications and interpretations of STS methodologies. The questions raised by the chapters are fundamental – philosophically, ethically and politically – and each, in their own way, reflect an orientation towards equity, care and justice. This includes, for example, Chapter 3's reconceptualization of health, Chapter 5's discussion of 'the artificial' in the context of AI, and Chapter 6's analysis of care teams. Each draws out how deeply imbricated assemblages of science and technology are with lived human and non-human relations and, in doing so, centres responsibility and solidarity as key ethical orientations.

I would like to close this introduction with some advice as it relates to how one might read these chapters. Specifically, I suggest that the reader embrace their differences in tone, focus and approach as consistent with the ethos of contemporary STS scholarship and, when finished, take a step back to reconceptualize and rethink what they have read in novel, imaginative ways. Together, these chapters offer a capacious means by which to reconstruct a health ecosystem that is care centred, participatory, situated and equitable.

References

Adams, C., Aydin, C., Blond, L., Funk, M., Ihde, D., Petersén, M. et al (2018) *Postphenomenological Methodologies: New Ways in Mediating Techno-Human Relationships*, Lanham: Rowman & Littlefield.

Bailey, M. and Peoples, W. (2017) 'Articulating black feminist health science studies', *Catalyst: Feminism, Theory, Technoscience*, 3(2): 1–27.

Barad, K. (2007) *Meeting the Universe Halfway: Quantum Physics and the Entanglement of Matter and Meaning*, Durham, NC: Duke University Press.

Bauchspies, W.K. and de La Bellacasa, M.P. (2009) 'Feminist science and technology studies: a patchwork of moving subjectivities. An interview with Geoffrey Bowker, Sandra Harding, Anne Marie Mol, Susan Leigh Star and Banu Subramaniam', *Subjectivity*, 28(1): 334–344.

Benezra, A. (2020) 'Race in the microbiome', *Science, Technology, & Human Values*, 45(5): 877–902.

Benjamin, R. (2016) 'Catching our breath: critical race STS and the carceral imagination', *Engaging Science, Technology, and Society*, 2: 145–156.

Bijker, W.E. and Law, J. (1992) *Shaping Technology/Building Society: Studies in Sociotechnical Change*, Cambridge, MA: MIT Press.

Blume, S., Galis, V. and Pineda, A.V. (2014) 'Introduction: STS and disability', *Science, Technology, & Human Values*, 39(1): 98–104.

Callon, M. and Law, J. (1997) 'Agency and the hybrid collectif', in B.H. Smith and A. Plotnitsky (eds) *Mathematics, Science, and Postclassical Theory*, Durham, NC: Duke University Press, pp 95–117.

Flore, J. (2021) 'Ingestible sensors, data, and pharmaceuticals: subjectivity in the era of digital mental health', *New Media & Society*, 23(7): 2034–2051.

Hackett, E.J., Amsterdamska, O., Lynch, M. and Wajcman, J. (2008) 'Introduction', in E.J. Hackett, O. Amsterdamska, M. Lynch and J. Wajcman (eds) *The Handbook of Science and Technology Studies* (3rd edn), London: MIT Press, pp 1–9.

Hatch, A.R. (2020) 'Du Boisian propaganda, Foucauldian genealogy, and antiracism in STS research', *Engaging Science, Technology, and Society*, 6: 58–65.

Henwood, F. and Marent, B. (2019) 'Understanding digital health: productive tensions at the intersection of sociology of health and science and technology studies', *Sociology of Health & Illness*, 41: 1–15.

Jasanoff, S. (2004) 'The idiom of co-production', in S. Jasanoff (ed) *States of Knowledge: The Co-production of Science and Social Order*, New York: Routledge, pp 1–12.

Jasanoff, S. and Kim, S.H. (2013) 'Sociotechnical imaginaries and national energy policies', *Science as Culture*, 22(2): 189–196.

Kolopenuk, J. (2020) 'Miskâsowin: Indigenous science, technology, and society', *Genealogy*, 4(1): 21.

Kuhn, T.S. (1962) *The Structure of Scientific Revolutions*, Chicago: University of Chicago Press.

Latour, B. (2000) 'When things strike back: a possible contribution of "science studies" to the social sciences', *The British Journal of Sociology*, 51(1): 107–123.

Latour, B. (2005) *Reassembling the Social: An Introduction to Actor-Network-Theory*, Oxford: Oxford University Press.

Latour, B. and Woolgar, S. (1979) *Laboratory Life: The Construction of Scientific Facts*, Princeton, NJ: Princeton University Press.

Law, J. (2004) *After Method: Mess in Social Science Research*, London: Routledge.

Lemke, T. (2015) 'New materialisms: Foucault and the "government of things"', *Theory, Culture & Society*, 32(4): 3–25.

Lupton, D. and Willis, K. (eds) (2021) *The COVID-19 Crisis: Social Perspectives*, London: Routledge.

Lyons, K., Parreñas, J.S., Tamarkin, N., Subramaniam, B., Green, L. and Pérez-Bustos, T. (2017) 'Engagements with decolonization and decoloniality in and at the interfaces of STS', *Catalyst: Feminism, Theory, Technoscience,* 3(1): 1–47.

Molldrem, S. and Thakor, M. (2017) 'Genealogies and futures of queer STS: issues in theory, method, and institutionalization', *Catalyst: Feminism, Theory, Technoscience,* 3(1): 1–15.

Oosterlaken, I., Grimshaw, D.J. and Janssen, P. (2012) 'Marrying the capability approach, appropriate technology and STS: the case of podcasting devices in Zimbabwe', in I. Oosterlaken and J. Hoven (eds) *The Capability Approach, Technology and Design*, Dordrecht: Springer, pp 113–133.

Pinch, T. and Leuenberger, C. (2006) 'Studying scientific controversy from the STS perspective', in *EASTS Conference Science Controversy and Democracy*, http://sts.nthu.edu.tw/easts/conference.htm.

Rudenko, N. (2016) 'Interfaces, efficiency, and inequality: the case of digital (auto-) ethnography of commercial technology', *International Journal of Actor-Network Theory and Technological Innovation (IJANTTI),* 8(4): 1–14.

Shotwell, A. (2016) *Against Purity: Living Ethically in Compromised Times*, Minneapolis: University of Minnesota Press.

Sikka, T. (2023) *Health Apps, Genetic Diets and Superfoods: When Biopolitics Meets Neoliberalism*, New York: Bloomsbury Publishing.

Sormani, P., Alač, M., Bovet, A. and Greiffenhagen, C. (2017) 'Ethnomethodology, video analysis, and STS', in U. Felt, R. Fouché, A.M. Clark and L. Laurel Smith-Doerr (eds) *The Handbook of Science and Technology Studies*, Cambridge, MA: MIT Press, pp 113–137.

Venturini, T., Munk, A.K. and Jacomy, M. (2019) 'Actor-network versus network analysis versus digital networks: are we talking about the same networks?' in J. Vertesi and D. Ribes (eds) *digitalSTS: A Field Guide for Science and Technology Studies,* Princeton, NJ: Princeton University Press, pp 510–524.

Vigren, M. and Bergroth, H. (2021) 'Move, eat, sleep, repeat: living by rhythm with proactive self-tracking technologies', *Nordicom Review,* 42(s4): 137–151.

1

Social and Behavioural Genomics and the Ethics of (In)Visibility

Daphne Oluwaseun Martschenko

Introduction

The postgenomic era – the period following the completion of the Human Genome Project in 2003 – ushered in rapid technological advancements and significantly reduced the costs of DNA sequencing. From genetic matchmaking services (Michels-Gualtieri and Appel, 2020) and in vitro fertilization for complex behaviours and traits (Orchid, nd), to direct-to-consumer genetic testing for behaviours and outcomes such as maths ability, depression and intelligence (GenePlaza, nd), genetic thinking is infiltrating countless aspects of public life. The growing accessibility of genomic data raises both possibilities (Harden, 2021a) and concerns (Roberts and Rollins, 2020) about the integration of molecular genetic data into society and public policy. It is transforming how individuals and communities think about themselves and others such that life is increasingly 'understood, and acted upon, at the molecular level' (Rose, 2007, p 5).

In the face of the DNA revolution, this chapter utilizes the theoretical framework of discriminate biopower (Fullwiley, 2004) to examine the forms of sociopolitical (in)visibility that may be crafted, upgraded or deleted in public education by contemporary research in social and behavioural genomics. Social and behavioural genomics examines how genetic differences between humans link to differences in behaviours and socioeconomic outcomes (Harden and Koellinger, 2020). While genetics have long captured the popular imagination, social and behavioural genomics is critically reimaging two aspects of our world today. First, how we relate our material, embodied selves, and second, the ways in which power and inequality are reflected and reproduced.

In building upon critical Science, Technology and Society scholarship on the relationship between genetics and society (for example, Duster, 2005; Bliss, 2013; Roberts, 2015; Panofsky, 2018), this chapter offers a critical interrogation of the sociotechnic imaginaries manufactured by social and behavioural genomics. It does so by exploring how potential applications of social and behavioural genomics to education might reaffirm or make new past, present and future (in)visibilities and inequalities, and feed into normative notions of who is 'healthy' and 'able'. These inequalities manifest themselves in the differential investment researchers, policy makers and societies give to support the health and wealth of individuals and collectives within public education. For example, this chapter explores how predictive applications of genomics to education research, policy and systems, including the idea of precision education, might map out sociotechnic futures in which students are tracked and streamed: (1) in the name of bio-molecular transformation; and (2) under the neoliberal guise of individualization, personalization and optimization. The infiltration of genomic data into education is birthing new potential strategies for 'governing' human life in schools and beyond. As such, I demonstrate how the creation, regulation, governance and marketing of genetics research is creating new strategies and contestations for human vitality that link the molecular, the population, the individual and the biological sciences together in complex ways.

Background and context

As molecular genetics research grows in prominence, some worry that holding a 'molecular gaze' has and will continue to threaten equity and justice (Bliss, 2018, 2012). Others argue that it is precisely what is needed to enhance social equality because it could help identify who is most vulnerable (and why) and enhance researchers' understandings of the effectiveness of public policies targeting equity (Harden, 2021b). This dual-use dilemma, in which genomics carries both risk and potential benefit, has long served as a steady source of debate. For instance, genetic ancestry testing has been used by White supremacists to validate their notions of racial purity (Panofsky and Donovan, 2019). Yet it has also empowered individuals to reject discrete racial categories and, for example, identify as multiracial on the US census (Johfre et al, 2021). Indigenous communities have pushed back on the inappropriate use of genetic ancestry testing to determine tribal affiliation (TallBear, 2013), while millions of individuals have taken at-home genetic ancestry tests (Regalado, 2019). And, while research into the genetic architecture of common diseases holds promise for illness prevention and early intervention (Visscher et al, 2021), there may be potentially negative psychosocial impacts of receiving a genetic test, such as lower perceived self-worth and confidence (Matthews et al, 2021).

Research into the genetics of human behaviour is especially rife with debate. Genetic ideologies have long been used to try to make sense of and address human behaviour and social phenomena such as inequalities in education and income. By and large, however, the history behind research in this area is troubling. For instance, the idea of discrete genetic differences in intelligence between racial groups was once used to resist the abolition of slavery (Evrie, 1868), outlaw interracial marriage (Pascoe, 2009), advocate for restricted immigration (Brigham, 1922) and fuel arguments that social policies can do little to secure equity in education outcomes (Jensen, 1969). Such efforts relied on distorted myths about biological differences between racial groups to divert attention away from the social forces that uphold racial classifications and racial paradigms about superiority and inferiority (Roberts and Rollins, 2020). In short, genetics have been and continue to be used to normalize racist, classist and ableist views about social stratification and inequality.

Today, research on the genetics of human behaviour continues in the field of social and behavioural genomics. Researchers study everything from fertility and reproductive behaviour (Mills et al, 2018) to same-sex sexual behaviour (Ganna et al, 2019), educational attainment (Lee et al, 2018) and household deprivation (Hill et al, 2016). Researchers in social and behavioural genomics are featured in prominent news outlets such as *The New Yorker* (Lewis-Kraus, 2021) and interviewed by the BBC (Rutherford, 2014). Some have argued that mainstream society and political progressives can no longer ignore the role genetics play in social outcomes and that understanding the relationship between genes and the environment is necessary for robust policy evaluation (Harden, 2021b) as well as for designing policies themselves (Behavioural Insights Team, 2019). More recently, some in the field have begun calling for anyone who claims to value egalitarianism, social justice or social equality to recognize how necessary genomic data is for realizing such aims (Harden, 2021b).

Social and behavioural genomics

To make sense of social and behavioural genomics and the arguments in support of or against it, one must first understand the field from which it arose: behaviour genetics. Behaviour genetics formally emerged in the 1950s as an attempt to establish a field interested in genetics and human behaviour that was 'free from the dark legacies of eugenics and scientific racism' (Nelson and Panofsky, 2018, p 291). Through twin studies, which compare similarities and differences in behaviours and outcomes between identical and fraternal twins, behaviour geneticists established the heritability of an array of human behaviours and began building their case for studying genetics further. Nevertheless, optimism about the viability of conducting socially neutral genetics research was soon thwarted by Arthur Jensen's 1969 publication 'How much can we boost IQ and scholastic achievement' (Jensen, 1969). In it, the American psychologist used behaviour genetics research to argue that intelligence, racial achievement gaps and social inequalities could not be resolved through social policies, since a person's genetics are fixed and play too big of a role in academic achievement (Nelson and Panofsky, 2018). Unfortunately for those in behaviour genetics, research on socially valued constructs such as intelligence remains highly contested, as 'the layer of theory between data and their interpretation is thicker and more opaque than in more established areas of science' (Turkheimer, 2015, p 32).

As DNA sequencing became more affordable with the completion of the Human Genome Project (National Human Genome Research Institute, 2016), researchers increasingly began using molecular genomic data to understand human behaviour. This shift from twin studies to genome-wide association studies (GWAS) led to the rise of social and behavioural *genomics*, which uses GWAS to examine large swaths of DNA and identify genetic variants of interest. Unlike behaviour genetics, which has strong roots in psychology and includes research pertaining to both animal and human genetics (Panofsky, 2014), social and behavioural genomics researchers tend to only be concerned with human genetics and come from a number of social science disciplines including economics, sociology and psychology. Social and behavioural genomics primarily seeks to integrate genomic data into the social sciences. Researchers argue that the social sciences could improve genetics research by enhancing understandings of the social dynamics that shape genetic disease (Shostak et al, 2009) and that genetics could improve the social sciences by providing additional data to conduct more rigorous empirical research (Harden, 2021b). In a more recent attempt to once again divorce genetics from determinism and essentialism, some researchers have begun to stipulate that social and behavioural genomics research could *enhance* egalitarianism and social justice efforts by reducing the blame placed on individuals for their circumstances (Harden, 2021b).

Given the array of arguments for and potential applications of social and behavioural genomics, it is unsurprising that the field has made particularly strong inroads in education. Researchers are investigating the genetic aetiologies of a range of education-related behaviours and outcomes, including intelligence (or cognitive ability) (Malanchini et al, 2020), educational attainment (Lee et al, 2018), ADHD (Ritter et al, 2017), dyslexia (Gialluisi et al, 2019), mathematics ability (Chen et al, 2017), reading ability (Luciano et al, 2013) and executive functioning (Engelhardt et al, 2016). Justifications for research in this area include the ability to enhance evidence-based educational interventions and support choice and equity in schools. For instance, one emerging argument is that understanding the role genetics play in differences between individuals could better inform whether educational interventions are effective and for whom (Harden and Koellinger, 2020). Genomic research on education-related behaviours and outcomes might also offer more robust ways to study and identify learning disabilities and improve understandings of how educational systems under- or over-diagnose students (Asbury, 2015; Kovas et al, 2016). Precision education, in which an individual student's genomic data is used to develop an individualized learning plan, could optimize education by helping to maximize a student's strengths and minimize their weaknesses, removing some of the pressures placed on overworked teachers in the process (Asbury and Plomin, 2013).

The rapid inroads social and behavioural genomic research is making into education research and settings is garnering the attention of policy makers, educators (Martschenko, 2019) and parents (Sabatello et al, 2021b; Au, 2022). In the US, K–12 education institutions are being approached by researchers with requests to genotype children (Hansen et al, 2015). At Yale, for instance, researchers partnered with New Haven Public Schools to assess the reading and cognitive abilities of first-graders in the hopes of creating a 'genetic screener for dyslexia' (New Haven Lexinome Project, nd). Additionally, a nationally representative survey of US parents found that many are interested in the use of genetic data to screen for learning disabilities (Sabatello et al, 2021b). In the United Kingdom, social and behavioural genomics researchers have appeared before the UK House of Commons Education Committee to discuss the role of genetics in shaping educational outcomes (House of Commons, 2014). Moreover, the Early Intervention Foundation, a UK nonprofit committed to improving the lives of youth 'at risk of experiencing poor outcomes', held a workshop series and published a 2021 report on the use of genetic data in early intervention and social policy (Asbury et al, 2021). In China, parents are conducting direct-to-consumer 'genetic talent testing' on their children in the hopes of identifying educational niches that may give their offspring a competitive edge (Au, 2022).

The American education system, which is the focus of this chapter, has historically used the language of genetics to assert, justify and reinscribe race- and class-based differences in academic achievement (Shockley, 1972; Jensen, 1991). It continues to be plagued by racial and socioeconomic disparities in a host of educational opportunities, including access to high-quality teachers (Goldhaber et al, 2015) and referral to gifted education (Martschenko, 2021b). I argue that the 'molecularization' (Fullwiley, 2008) of educational outcomes such as educational attainment contributes to pre-existing systems of power in critical ways that are informed by our tendency to think of genes and genetics in deterministic and fatalistic terms (Heine et al, 2017) and to strategically employ this determinism in service of our pre-existing beliefs and values (Condit, 2019).

Discriminate biopower

The origin of the biopolitical analysis employed in this chapter is grounded in the works of French philosophical theorist Michel Foucault. Biopower is 'the set of mechanisms through which the basic biological features of the human species became the object of a political strategy' (Foucault, 2007, p 1). Biopower entails:

(1) One or more truth discourses about the 'vital' character of living human beings.
(2) An array of authorities considered competent to speak that truth.
(3) Strategies for intervention upon collective existence in the name of life and health.
(4) Modes of subjectivation, in which individuals work on themselves in the name of individual or collective life or health. (Rabinow and Rose, 2006, p 197)

Through subtle regulations, expectations, and norms that are encoded into social practices and human behaviour (Foucault, 1978), biopower operates as a set of events, discourses or relations that institutionalize routines and normalize interventions that optimize certain forms of life over others (McWhorter, 2004; Anderson, 2012). For instance, the deluge of genetic information in the form of genetic research, screening and direct-to-consumer testing is building expectations that individuals 'draw on science to articulate their own judgments and political claims' and 'take greater responsibility for their own vitality' (Raman and Tutton, 2009, p 6). There is great interest in using genomic data to 'recognize individuals at risk and the type of risk individuals incur throughout their life' (Foucault and Senellart, 2008, p 227).

Discriminate biopower examines the differential investments individuals and communities are given on the basis of categories such as race, class, gender, (dis)ability, citizenship and language (Martschenko, 2020). Anthropologist Duana Fullwiley first used the term to describe 'the utter patchiness of what Foucault depicts as ordered "interventions and regulatory controls"' (Fullwiley, 2004, p 160). In her work on sickle-cell in Senegal, Fullwiley demonstrates the 'socio-political invisibility' experienced by those whose health and illness is entertained by medical experts with high levels of reluctance, if ever (Fullwiley, 2004, p 159). The 'socio-political invisibility' experienced by those with sickle-cell – a racialized disease that for a long time remained 'clinically invisible' (Wailoo, 2017, p 805) – forces individuals and communities into the struggle to secure their own vitality with limited institutional investment; it is the by-product of a biopower that is 'uneven and variable in its distribution and attention' (Fullwiley, 2004, p 160).

In education, discriminate biopower offers a theoretical lens through which to understand how educational inequalities might be enhanced and regularized through 'molecularized' sciences like social and behavioural genomics. It acknowledges the numerous factors (for example, race, class, ability) that come together to weave an intricate web of advantage, disadvantage and everything in between. Discriminate biopower is concerned with how biopolitical states invest in bodies and which bodies they choose to empower. This framework raises questions such as: who gets to define, distribute, amass and acquire biopolitical investment? In the US education system, these investment practices manifest in the painful reality that some students will receive better educational opportunities, resources and services than others. Such realities are strongly rooted in racism and the legacies of slavery, reinforced by efforts to resist desegregation (Porter, 2017) and maintained through neoliberal education policies (Brathwaite, 2016). Discriminate biopower not only builds an understanding of social and behavioural genomics, and the implications of this research for education; it contextualizes discriminatory education practices that genetics research might normalize. As such, this theoretical framework enables an exploration of how scientific and educational institutions interact to maintain themselves and the systems in which they operate.

The sociotechnic imaginary

Employing discriminate biopower, I document the sociotechnic imaginaries that arise from the integration of social and behavioural genomics into education. The sociotechnic imaginary is 'the collectively held and performed visions of desirable futures ... animated by shared understandings of forms of social life and social order attainable through, and supportive of, advances in science and technology' (Jasanoff and Kim, 2015, p 25). This imaginary resides at the social rather than the individual level and generates shared systems of meaning and belonging, and collective visions of the world and how it is structured (Jasanoff and Kim, 2015). Research on the genetics of human behaviour crafts sociotechnic imaginaries that employ 'genomic authority' and the 'allure of objectivity' (Benjamin, 2015). Although Jasanoff and Kim (2015) define the sociotechnic imaginary as evoking 'desirable futures', I acknowledge that a future that is desirable for some may not be for others. For example, current genomic research on intelligence could be used to normatively determine not only who is more or less able, but who is more valuable to society, thereby (re)imagining social stratification. In a world where policies are constructed to favour those considered 'more valuable' and/or a world in which individuals are treated differently on the basis of whether their genetics are considered favourable or not, it is likely that a few will benefit at the expense of the many. On the other hand, in an alternate universe, genetics could become an 'antidote to blame' and used to advocate for the redistribution of resources to support the flourishing of those who are 'at risk' (Harden, 2021b, p 195). In practice, however, I argue that it may be hard to create a 'desirable future' that is desirable for all.

Society is fascinated by genetic research on behaviour 'because it leads to the edge of the possible' (Turkheimer, 2015, p 38). The powerful hold genes have on the popular imagination (Lee et al, 2008), coupled with the growing marketization of genetic technologies, is reconstructing social relations and cultural meanings inside and outside of education. Current debates over whether and how to integrate molecular genetic data into education research and policy therefore raise two critical questions: (1) what sociotechnic imaginaries are created by social and behavioural genomics; and (2) 'who and what are fixed in place – classified, corralled, and/or coerced – to enable technoscientific development'? (Benjamin, 2016, p 146). In other words, as social and behavioural genomics strengthen their grip on the popular imagination, what truth discourses does the field create and promote? And, who benefits in these emerging imaginaries? Answering these questions through a discriminate biopower framework highlights the uneven and variable ways in which vital characteristics of human existence, such as health and wealth (Rabinow and Rose, 2006), are applied to individuals and communities in educational settings.

The sociotechnic imaginary of technological upgrades

Can genetics offer a 'technological fix' (Johnston, 2018) for problems traditionally considered to be socially, politically or culturally derived? Or at least improve our understanding of these issues? Sociotechnic imaginaries express the elements of a 'good life' and are used to inform and legitimize innovation policies (Levidow and Raman, 2020). The imaginary offered up by enthusiasts of social and behavioural genomics includes integrating genomic data into the social sciences to identify previously hidden social influences (Herd et al, 2021) and using genomics to address the 'lack of knowledge' and 'paucity of data' about what works and does not work in education (Harden, 2021b, p 176). For instance, using genomic predictive technologies, learning disabilities might be identified earlier and more often (Shero et al, 2021) and students who may have previously fallen through the cracks due to, for example, undiagnosed dyslexia, could receive greater attention when genomic data is used to shed light on who may struggle in school and why. Tapping into existing neoliberal logics, the sociotechnic imaginary also anticipates that with the acceptance of the role of genetics, students will be given the opportunity to more freely pursue their passions through increased choice, and teachers will be given the liberty to be more flexible in their teaching (Asbury and Plomin, 2013). If this were the case, personalized, student-centred learning would take on a new meaning.

Questions of egalitarianism and racial justice are central to the social and behavioural genomics fairy tale; society should be structured to benefit all regardless of genetics but needs to be aware about genetics to do so (Harden, 2021b, p 94). Therefore, and in the spirit of transformational possibility, those of sound mind are called upon to lift their heads up and out of the sand. 'Socially identified authorities', or researchers (Rabinow and Rose, 2003, p 4), normalize ways of thought and understanding about the biology of individual differences. The dominant truth discourse is that genetics 'tell your fortune from the moment of your birth' and are 'completely reliable and unbiased' (Plomin, 2018, p vii); for this reason DNA matters and genomic research can and should be applied to education research, policy, policy evaluation and, in some instances, practice. As this discourse gains political power, it could influence how individuals understand and organize themselves, how students are taught and how educators teach. It promises to offer much-needed answers to long-standing questions about what makes students differ in their educational and life-long trajectories and what can be done about it.

If embracing genetics can make visible what was once invisible, and if genetics can act as an 'antidote to blame' (Harden, 2021b, p 195), then policy makers should be ready to recommend redistributive policies and the public should be more open to embracing them. Differential investment would then be provided on the basis of 'risk' for lower educational attainment, ADHD or dyslexia. In this imaginary, the 'uneven and variable' distribution of resources and attention (Fullwiley, 2004, p 160) appears to upgrade rather than delete the sociopolitical visibility of those relegated to the peripheries; 'protecting the most vulnerable requires knowing who is most vulnerable, identifying what factors make them most vulnerable, and structuring society for their benefit' (Harden, 2021b, p 255). In other words, genetics shed light on previously invisible vulnerabilities so that society might respond.

The sociotechnic imaginary of technological deletions

The same idea that undergirds positive sociotechnic imaginaries about genomics (that genetics are reliable and unbiased and therefore genomic research can and should be applied to education research, policy, policy evaluation and even practice) comprises the very foundation of a different kind of imaginary. Instead of employing selectivity to equitably *re*distribute resources, it is used to inequitably concentrate them. Rendering disparate educational outcomes a product of individual biology might therefore distract from structural and historical determinants (Martschenko, 2021). What if genomics benefit some and not all? Or serve to perpetuate the long-standing tradition of dividing people into discrete groups? What if genomics produces forms of life that *actually* create or reinscribe inequitable differences in the body and its functions through technologies like in vitro fertilization and human germline editing? Who gets to decide what the 'optimal' state of being is? On the flip side of the redistributive, egalitarian genomic imaginary is one that engenders 'social and spatial marginalization' (Wacquant, 2007, p 72), resuscitates mythologies about inherent differences between racial groups and validates the ongoing sociopolitical invisibility of those cast to the periphery.

Society systematically stratifies individuals and groups to receive different amounts of investment. In genomics, this stratification results in research prioritizations that ignore large swaths of the global population and reflect colonial logics. Genomic studies overwhelmingly recruit individuals of European genetic ancestries (Popejoy and Fullerton, 2016), meaning that the findings from these genetic studies, and any potential benefits they may afford, are confined to a narrow subset of the global population (Martin et al, 2017). Additionally, populations that are a mix of genetic ancestries are often discarded from analysis altogether, due to their biological complexity (Bustamante et al, 2011). Finally, ambiguous and even problematic population classification schema heighten cultural obsessions with racial difference, positioning genetic ancestry as emblematic of race (Panofsky and Bliss, 2017; Popejoy, 2021). Taken together, these phenomena illustrate how the sociopolitical invisibility of those who fall outside Western or European categories is maintained through genomics research.

In education, individuals are frequently othered via the creation of hierarchized and binary 'sets' (for example, the educated and uneducated, English-language learners and native speakers, citizens and immigrants) (Weheliye, 2014). Some of these 'sets' are 'classified as deviating from full … humans' (Weheliye, 2014, p 60). Genomics threaten to reimagine stratification in education as symbolic 'evidence' of a biological pecking order; as easily as genomics might be used for redistributive justice, it can be used to legitimize inequitable 'modes of investment' that rely on arbitrary definitions of the worthy and unworthy (Foucault and Senellart, 2008, p 77). For instance, ability grouping is commonly used in the US education system to sort students and identify them for socially coveted programmes like gifted and talented education. If genetics were incorporated into screening and identification practices, those designated to the 'extremes' of ability might find themselves in special education or gifted education programmes even before they exhibited 'symptoms.' Additionally, schools might decide to deny admissions to students at risk for a host of costly developmental disorders (Martschenko et al, 2019). For the individual, receiving genetic test results for an outcome such as educational attainment could result in lower levels of confidence or perceived self-worth (Matthews et al, 2021) and fuel stigma

(Sabatello, 2018). In this way, genetic testing could emerge as a new iteration of the high-stakes testing used to craft accountability narratives within the larger neoliberal project of educational reform (Au, 2016).

More often than not, it is teachers who are key decision makers in students' educational futures. Many disproportionately place students of colour in lower-ranked groups (Gillborn, 2010) or under-identify them for gifted education services (Martschenko, 2021). Educators' views of their students inform students' perceptions of themselves and the preparation they receive for higher education (Blanchard and Muller, 2015). The critical role of teachers within education has led some scholars to explore the relationship between social and behavioural genomics research and teacher perceptions of student ability and achievement (Crosswaite and Asbury, 2018; Martschenko, 2019). Furthermore, a burgeoning body of literature is considering the potential ramifications of genetics for equity, access, inclusion and justice in education (Sabatello, 2018; Martschenko et al, 2019; Sabatello et al, 2021a). These developments illustrate how the integration of molecular genetic data into education could upset the already sparse attention given to initiatives aimed at promoting mobility and equity.

Looking ahead

The future of social and behavioural genomics in education will likely fall somewhere in between the sociotechnic imaginaries of technological upgrades and deletions. Nevertheless, each of these imaginaries offers ways to reimagine social and economic opportunities through a molecular gaze. In one imaginary, genetics are used to redistribute time, energy, money and resources in service of educational equity and social justice. In the other, genomics are used to accrue or 'hoard' potential benefits among those who are already privileged. In both, the dark legacy of eugenics looms large.

Importantly, in order for social and behavioural genomics to be of practical use to educators and education systems, measures of a genetic effect will have to be suitably large and practically measurable. Yet a genetic effect that is both large and measurable enough to be used introduces the risk of determinism, essentialism, classism or racism in its application. Genetic ideologies have long permeated attitudes towards difference and ability and, by extension, intrinsic value and worth. There is a very fine line between what could be helpful and what could be harmful.

Conclusion

Genetic discourses and technologies have real implications for education systems and the stakeholders and participants within them. Discussion of the relationship between genetics and educational outcomes is not new. What is new, is *how* social and behavioural genomics research is converging with systems of education to produce sociotechnic imaginaries that either enhance or further diminish the sociopolitical visibility of marginalized groups. While previous research made biological arguments without any genetic data to support it (for example, Herrnstein and Murray, 1996), today's researchers are using molecular genetic data to study education-related behaviours and outcomes. The floodgates of genetic data are wide open. What does this mean for education?

Education is frequently proposed as a gateway to social mobility, but in the US education suffers from racial and socioeconomic disparities in school quality (Berkowitz et al, 2016), educational attainment and test-scores (Reardon, 2016), and rates of discipline (Carter et al, 2017). Public perceptions of intelligence or ability, which are rooted in an ugly history (Roberts, 2015), also shape a person's life trajectory. In much of the Western world, being labelled 'intelligent' or 'gifted' signifies, to some degree, what one's future will look like. Students are deemed 'intelligent' if their grades are outstanding, coursework rigorous, test scores competitive and college prospects high. Intelligence, social and behavioural genomics says, 'spills over into many aspects of everyday life' (Plomin and Stumm, 2018, p 1) and helps to explain differences in school performance 'which in turn lead to social and economic opportunities such as those related to occupation and income' (Plomin and Stumm, 2018, p 1). Social and behavioural genomics introduces forms of molecularization that influence our thinking, inform our interactions with others and may potentially shape our public policies.

The imaginaries held by social and behavioural genomics are in tension with each other. This tension speaks to the contested nature of the research and the dual-use dilemma that studying the genetics of human behaviour lives with. As genomics spill over into everyday life, a particular truth discourse is solidifying: that genetics have the authority to tell us about human behaviour and that our policies should be 'predicated on genomic authority' (Benjamin, 2015, p 140). Unless we engage in a critical dialogue on what can, should and needs to be done about social and behavioural genomics in education, racist, classist and inequitable imaginaries and the forms of (in)visibility they engender will take root in an environment that already regularizes them.

References

Anderson, B. (2012) 'Affect and biopower: towards a politics of life', *Transactions Institute of British Geographers*, 37: 28–43, https://doi.org/10.1111/j.1475-5661.2011.00441.x.

Asbury, K. (2015) 'Can genetics research benefit educational interventions for all?' *Hastings Center Report*, 45(5Suppl): S39–42.

Asbury, K. and Plomin, R. (2013) *G Is for Genes: The Impact of Genetics on Education and Achievement, Understanding Children's Worlds,* Chichester: Wiley-Blackwell.

Asbury, K., McBride, T. and Rimfeld, K. (2021) *Genetics and Early Intervention: Exploring Ethical and Policy Questions,* London: Early Intervention Foundation.

Au, L. (2022) 'Testing the talented child: direct-to-consumer genetic talent tests in China', *Public Understanding of Science,* 31: 195–210, https://doi.org/10.1177/09636625211051964.

Au, W. (2016) 'Meritocracy 2.0: high-stakes, standardized testing as a racial project of neoliberal multiculturalism', *Education Policy,* 30: 39–62, https://doi.org/10.1177/0895904815614916.

Behavioural Insights Team (2019) 'Informing policy decisions with evidence from behavioural genetics with Robert Plomin', https://www.bi.team/informing-policy-decisions-with-evidence-from-behavioural-genetics-with-robert-plomin/.

Benjamin, R. (2015) 'The emperor's new genes: science, public policy, and the allure of objectivity', *Annals of the American Academy of Political and Social Science,* 661: 130–142, https://doi.org/10.1177/0002716215587859.

Benjamin, R. (2016) 'Catching our breath: critical race STS and the carceral imagination', *Engaging Science, Technology, and Society,* 2: 145–156, https://doi.org/10.17351/ests2016.70.

Berkowitz, R., Moore, H., Astor, R.A. and Benbenishty, R. (2016) 'A research synthesis of the associations between socioeconomic background, inequality, school climate, and academic achievement', *Review of Educational Research* 87(2), https://10.3102/0034654316669821.

Blanchard, S. and Muller, C. (2015) 'Gatekeepers of the American dream: how teachers' perceptions shape the academic outcomes of immigrant and language-minority students', *Social Science Research,* 51: 262–275, https://doi.org/10.1016/j.ssresearch.2014.10.003.

Bliss, C. (2012) *Race Decoded: The Genomic Fight for Social Justice,* Stanford, CA: Stanford University Press.

Bliss, C. (2013) 'The marketization of identity politics', *Sociology* 47: 1011–1025, https://doi.org/10.1177/0038038513495604.

Bliss, C. (2018) *Social by Nature: The Promise and Peril of Sociogenomics,* Stanford, CA: Stanford University Press.

Brathwaite, J. (2016) 'Neoliberal education reform and the perpetuation of inequality', *Critical Sociology,* 43(3), https://doi.org/10.1177/0896920516649418.

Brigham, C.C. (1922) *A Study of American Intelligence,* Princeton, NJ: Princeton University Press.

Bustamante, C.D., De La Vega, F.M. and Burchard, E.G. (2011) 'Genomics for the world', *Nature,* 475: 163–165, https://doi.org/10.1038/475163a.

Carter, P.L., Skiba, R., Arredondo, M.I. and Pollock, M. (2017) 'You can't fix what you don't look at: acknowledging race in addressing racial discipline disparities', *Urban Education*, 52: 207–235, https://doi.org/10.1177/0042085916660350.

Chen, H., Gu, X., Zhou, Y., Ge, Z., Wang, B., Siok, W.T. et al (2017) 'A genome-wide association study identifies genetic variants associated with mathematics ability', *Scientific Reports*, 7: 40365, https://doi.org/10.1038/srep40365.

Condit, C.M. (2019) 'Laypeople are strategic essentialists, not genetic essentialists', *Hastings Center Report,* 49: S27–S37, https://doi.org/10.1002/hast.1014.

Crosswaite, M. and Asbury, K. (2018) 'Teacher beliefs about the aetiology of individual differences in cognitive ability, and the relevance of behavioural genetics to education', *British Journal of Educational Psychology*, https://doi.org/10.1111/bjep.12224.

Duster, T. (2005) 'Race and reification in science', *Science,* 307: 1050–1051, https://doi.org/10.1126/science.1110303.

Engelhardt, L.E., Mann, F.D., Briley, D.A., Church, J.A., Harden, K.P. and Tucker-Drob, E.M. (2016) 'Strong genetic overlap between executive functions and intelligence', *Journal of Experimental Psychology: General,* 145: 1141–1159, https://doi.org/10.1037/xge0000195.

Evrie, J.H.V. (1868) *Negroes and Negro Slavery: The First an Inferior Race: The Latter Its Normal Condition*, Horton, NY: Van Evrie.

Foucault, M. (2007) *Security, Territory, Population: Lectures at the College De France, 1977–78*, New York: Springer.

Foucault, M. and Senellart, M. (2008) *The Birth of Biopolitics: Lectures at the Collège de France, 1978–79*, Basingstoke: Palgrave Macmillan.

Fullwiley, D. (2004) 'Discriminate biopower and everyday biopolitics: views on sickle cell testing in Dakar', *Medical Anthropology,* 23: 157–194, https://doi.org/10.1080/01459740490448939.

Fullwiley, D. (2008) 'The molecularization of race and institutions of difference: pharmacy and public science after the genome', in B.A. Koenig, S.S. Lee and S. Richardson (eds) *Revisiting Race in a Genomic Age (Studies in Medical Anthropology)*, New Brunswick, NJ: Rutgers University Press, pp 149–171.

Ganna, A., Verweij, K., Nivard, M., Maier, R., Wedow, R., Busch, A. et al (2019) 'Large-scale GWAS reveals insights into the genetic architecture of same-sex sexual behavior', *Science* 365, https://doi.org/10.1126/science.aat7693.

GenePlaza (nd) *GenePlaza | App Store – Intelligence App*, www.geneplaza.com.

Gialluisi, A., Andlauer, T.F.M., Mirza-Schreiber, N., Moll, K., Becker, J., Hoffmann, P. et al (2019) 'Genome-wide association scan identifies new variants associated with a cognitive predictor of dyslexia', *Translational Psychiatry* 9: 1–15, https://doi.org/10.1038/s41398-019-0402-0.

Gillborn, D. (2010) 'Reform, racism and the centrality of whiteness: assessment, ability and the "new eugenics"', *Irish Educational Studies* 29: 231–252, https://doi.org/10.1080/03323315.2010.498280.

Goldhaber, D., Lavery, L. and Theobald, R. (2015) 'Uneven playing field? Assessing the teacher quality gap between advantaged and disadvantaged students', *Education Research*, 44: 293–307, https://doi.org/10.3102/00131 89X15592622.

Hansen, E.T., Gluck, S. and Shelton, A.L. (2015) 'Obligations and concerns of an organization like the Center for Talented Youth', *Hastings Center Report,* 45: S66–S72, https://doi.org/10.1002/hast.502.

Harden, K.P. (2021a) '"Reports of my death were greatly exaggerated": behavior genetics in the postgenomic era', *Annual Review of Psychology,* 72, https://doi.org/10.1146/annurev-psych-052220-103822.

Harden, K.P. (2021b) *The Genetic Lottery: Why DNA Matters for Social Equality*, Princeton, NJ: Princeton University Press.

Harden, K.P. and Koellinger, P.D. (2020) 'Using genetics for social science', *Nature Human Behaviour*, 4: 567–576, https://doi.org/10.1038/s41 562-020-0862-5.

Heine, S.J., Dar-Nimrod, I., Cheung, B.Y. and Proulx, T. (2017) 'Essentially biased: why people are fatalistic about genes', in J.M. Olson (ed) *Advances in Experimental Social Psychology*, New York: Academic Press, pp 137–192.

Herd, P., Mills, M.C. and Dowd, J.B. (2021) 'Reconstructing sociogenomics research: dismantling biological race and genetic essentialism narratives', *Journal of Health and Social Behavior,* 62: 419–435, https://doi.org/10.1177/00221465211018682.

Herrnstein, R.J. and Murray, C.A. (1996) *The Bell Curve: Intelligence and Class Structure in American Life*, New York: Simon & Schuster.

Hill, W.D., Hagenaars, S.P., Marioni, R.E., Harris, S.E., Liewald, D.C.M., Davies, G. et al (2016) 'Molecular genetic contributions to social deprivation and household income in UK biobank', *Current Biology,* https://doi.org/10.1016/j.cub.2016.09.035.

House of Commons Education Committee (2014) *Underachievement in Education by White Working Class Children*, First Report of Session 2014–15.

Jasanoff, S. and Kim, S.-H. (eds) (2015) *Dreamscapes of Modernity: Sociotechnical Imaginaries and the Fabrication of Power*, Chicago: University of Chicago Press.

Jensen, A.R. (1969) 'How much can we boost IQ and scholastic achievement', *Harvard Education Review,* 39: 1–123, https://doi.org/10.17763/haer.39.1.l3u15956627424k7.

Jensen, A.R. (1991) 'Spearman's "g" and the problem of educational equality', *Oxford Review of Education,* 17: 169–187.

Johfre, S.S., Saperstein, A. and Hollenbach, J.A. (2021) 'Measuring race and ancestry in the age of genetic testing', *Demography,* 58: 785–810, https://doi.org/10.1215/00703370-9142013.

Johnston, S.F. (2018) 'Alvin Weinberg and the promotion of the technological fix', *Technology and Culture,* 59: 620–651, https://doi.org/10.1353/tech.2018.0061.

Kovas, Y., Tikhomirova, T., Selita, F., Tosto, M.G. and Malykh, S. (2016) 'How genetics can help education', in Y. Kovas, S. Malykh and D. Gaysina (eds) *Behavioural Genetics for Education,* Basingstoke: Palgrave Macmillan, pp 1–23.

Lee, J.J., Wedow, R., Okbay, A., Kong, E., Maghzian, O., Zacher, M. et al (2018) 'Gene discovery and polygenic prediction from a genome-wide association study of educational attainment in 1.1 million individuals', *Nature Genetics,* 1, https://doi.org/10.1038/s41588-018-0147-3.

Lee, S.S.-J., Mountain, J., Koenig, B., Altman, R., Brown, M., Camarillo, A. et al (2008) 'The ethics of characterizing difference: guiding principles on using racial categories in human genetics', *Genome Biology,* 9: 1–4, https://doi.org/10.1186/gb-2008-9-7-404.

Levidow, L. and Raman, S. (2020) 'Sociotechnical imaginaries of low-carbon waste-energy futures: UK techno-market fixes displacing public accountability', *Social Studies of Science,* 50: 609–641, https://doi.org/10.1177/0306312720905084.

Lewis-Kraus, G. (2021) 'Can progressives be convinced that genetics matters?', *The New Yorker,* 13 September.

Luciano, M., Evans, D.M., Hansell, N.K., Medland, S.E., Montgomery, G.W., Martin, N.G. et al (2013) 'A genome-wide association study for reading and language abilities in two population cohorts', *Genes, Brain and Behavior,* 12: 645–652, https://doi.org/10.1111/gbb.12053.

Malanchini, M., Rimfeld, K., Allegrini, A.G., Ritchie, S.J. and Plomin, R. (2020) 'Cognitive ability and education: how behavioural genetic research has advanced our knowledge and understanding of their association', *Neuroscience and Biobehavioral Reviews,* 111: 229–245, https://doi.org/10.1016/j.neubiorev.2020.01.016.

Martin, A.R., Gignoux, C.R., Walters, R.K., Wojcik, G.L., Neale, B.M., Gravel, S. et al (2017) 'Human demographic history impacts genetic risk prediction across diverse populations', *American Journal of Human Genetics,* 100: 635–649, https://doi.org/10.1016/j.ajhg.2017.03.004.

Martschenko, D. (2019) '"DNA dreams": teacher perspectives on the role and relevance of genetics for education', *Research in Education,* 107(1): 1–22, https://doi.org/10.1177/0034523719869956.

Martschenko, D. (2020) 'Embodying biopolitically discriminate borders: teachers' spatializations of race', *Discourse: Studies in the Cultural Politics of Education,* 43(1): 101–114, https://doi.org/10.1080/01596306.2020.1813089.

Martschenko, D. (2021) 'Normalizing race in (gifted) education: genomics and spaces of White exceptionalism', *Critical Studies in Education,* 64(1): 67–83, https://doi.org/10.1080/17508487.2021.1978517.

Martschenko, D., Trejo, S. and Domingue, B.W. (2019) 'Genetics and education: recent developments in the context of an ugly history and an uncertain future', *AERA Open,* 5: 1–15, https://doi.org/10.1177/23328 58418810516.

Matthews, L.J., Lebowitz, M.S., Ottman, R. and Appelbaum, P.S. (2021) 'Pygmalion in the genes? On the potentially negative impacts of polygenic scores for educational attainment', *Social Psychology of Education,* 24: 789–808, https://doi.org/10.1007/s11218-021-09632-z.

McWhorter, L. (2004) 'Sex, race, and biopower: a Foucauldian genealogy', *Hypatia,* 19: 38–62, https://doi.org/10.1111/j.1527-2001.2004.tb01301.x.

Michels-Gualtieri, M. and Appel, J.M. (2020) 'The illusion of genetic romance', *Scientific American Blog Network,* https://blogs.scientificameri can.com/observations/the-illusion-of-genetic-romance/.

Mills, M.C., Barban, N. and Tropf, F.C. (2018) 'The sociogenomics of polygenic scores of reproductive behavior and their relationship to other fertility traits', *RSF: The Russell Sage Foundation Journal of the Social Sciences,* 4: 122–136, https://doi.org/10.7758/rsf.2018.4.4.07.

National Human Genome Research Institute (2016) 'The cost of sequencing a human genome', www.genome.gov/27565109/The-Cost-of-Sequenc ing-a-Human-Genome.

Nelson, N.C. and Panofsky, A. (2018) 'Behavior genetics: boundary crossings and epistemic cultures', in S. Gibbon, B. Prainsack, S. Hilgartner and J. Lamoreaux (eds) *Routledge Handbook of Genomics, Health and Society,* Abingdon: Routledge.

New Haven Lexinome Project (nd) http://yalenhlp.org/.

Orchid (nd) 'Have healthy babies', www.orchidhealth.com/.

Panofsky, A. (2014) *Misbehaving Science: Controversy and the Development of Behavior Genetics,* Chicago: University of Chicago Press.

Panofsky, A. (2018) 'Rethinking scientific authority: behavior genetics and race controversies', *American Journal of Cultural Sociology,* 6: 322–358, https://doi.org/10.1057/s41290-017-0032-z.

Panofsky, A. and Bliss, C. (2017) 'Ambiguity and scientific authority: population classification in genomic science', *American Sociological Review,* 82: 59–87, https://doi.org/10.1177/0003122416685812.

Panofsky, A. and Donovan, J. (2019) 'Genetic ancestry testing among white nationalists: from identity repair to citizen science', *Social Studies of Science,* 49: 653–681, https://doi.org/10.1177/0306312719861434.

Pascoe, P. (2009) *What Comes Naturally: Miscegenation Law and the Making of Race in America*, Oxford: Oxford University Press.

Plomin, R. (2018) *Blueprint: How DNA Makes Us Who We Are*, Cambridge, MA: MIT Press.

Plomin, R. and Stumm, S. (2018) 'The new genetics of intelligence', *Nature Reviews Genetics*, 19: 148–159, https://doi.org/10.1038/nrg.2017.104.

Popejoy, A.B. (2021) 'Too many scientists still say Caucasian', *Nature*, 596: 463, https://doi.org/10.1038/d41586-021-02288-x.

Popejoy, A.B. and Fullerton, S.M. (2016) 'Genomics is failing on diversity', *Nature News*, 538: 161, https://doi.org/10.1038/538161a.

Porter, J.W. (2017) 'A "precious minority": constructing the "gifted" and "academically talented" student in the era of Brown v Board of Education and the National Defense Education Act', *Isis*, 108: 581–605, https://doi.org/10.1086/694446.

Rabinow, P. and Rose, N. (2003) 'Thoughts on the concept of biopower today', *BioSocieties*, 1: 195–217.

Raman, S. and Tutton, R. (2009) 'Life, science, and biopower', *Science, Technology, & Human Values*, 35(5): 711–734, https://doi.org/10.1177/0162243909345838.

Reardon, S.F. (2016) 'School segregation and racial academic achievement gaps', *RSF: The Russell Sage Foundation Journal of the Social Sciences*, 2: 34–57, https://doi.org/10.7758/RSF.2016.2.5.03.

Regalado, A. (2019) 'More than 26 million people have taken an at-home ancestry test', *MIT Technology Review*, February.

Ritter, M.L., Guo, W., Samuels, J.F., Wang, Y., Nestadt, P.S., Krasnow, J. et al (2017) 'Genome wide association study (GWAS) between attention deficit hyperactivity disorder (ADHD) and obsessive compulsive disorder (OCD)', *Frontiers of Molecular Neuroscience*, 10, https://doi.org/10.3389/fnmol.2017.00083.

Roberts, D. (2015) 'Can research on the genetics of intelligence be "socially neutral"?' *Hastings Center Report*, 45: S50–S53, https://doi.org/10.1002/hast.499.

Roberts, D. and Rollins, O. (2020) 'Why sociology matters to race and biosocial science', *Annual Review of Sociology*, 46: 195–214, https://doi.org/10.1146/annurev-soc-121919-054903.

Rose, N. (2007) *The Politics of Life Itself: Biomedicine, Power, and Subjectivity in the Twenty-First Century*, Princeton, NJ: Princeton University Press.

Rutherford, A. (2014) 'Born smart' [podcast], Intelligence: Born Smart, Born Equal, Born Different, episode 1, BBC Radio 4, www.bbc.co.uk/programmes/b042q944/episodes/downloads.

Sabatello, M. (2018) 'A genomically informed education system? Challenges for behavioral genetics', *Journal of Law, Medicine & Ethics*, 46: 130–144, https://doi.org/10.1177/1073110518766027.

Sabatello, M., Insel, B.J., Corbeil, T., Link, B.G. and Appelbaum, P.S. (2021a) 'The double helix at school: behavioral genetics, disability, and precision education', *Social Science & Medicine,* 278: 113924, https://doi.org/10.1016/j.socscimed.2021.113924.

Sabatello, M., Martin, B., Corbeil, T., Lee, S., Link, B.G. and Appelbaum, P.S. (2021b) 'Nature vs. nurture in precision education: insights of parents and the public', *AJOB Empirical Bioethics,* https://doi.org/10.1080/23294515.2021.1983666.

Shero, J., van Dijk, W., Edwards, A., Schatschneider, C., Solari, E.J. and Hart, S.A. (2021) 'The practical utility of genetic screening in school settings', *Npj Science of Learning,* 6: 1–10, https://doi.org/10.1038/s41539-021-00090-y.

Shockley, W. (1972) 'Dysgenics, geneticity, raceology: a challenge to the intellectual responsibility of educators', *Phi Delta Kappan,* 53: 297–307.

Shostak, S., Freese, J., Link, B.G. and Phelan, J.C. (2009) 'The politics of the gene: social status and beliefs about genetics for individual outcomes', *Social Psychology Quarterly,* 72: 77–93.

TallBear, K. (2013) 'Native American DNA: tribal belonging and the false promise of genetic science', https://openlibrary.org/books/OL26181069M.

Turkheimer, E. (2015) 'Genetic prediction', *Hastings Center Report,* 45: S32–S38, https://doi.org/10.1002/hast.496.

Visscher, P.M., Yengo, L., Cox, N.J. and Wray, N.R. (2021) 'Discovery and implications of polygenicity of common diseases', *Science,* 373: 1468–1473, https://doi.org/10.1126/science.abi8206.

Wacquant, L. (2007) 'Territorial stigmatization in the age of advanced marginality', *Thesis Eleven,* 91: 66–77, https://doi.org/10.1177/0725513607082003.

Wailoo, K. (2017) 'Sickle cell disease: a history of progress and peril', *New England Journal of Medicine,* 376: 805–807, https://doi.org/10.1056/NEJMp1700101.

Weheliye, A.G. (2014) *Habeas Viscus: Racializing Assemblages, Biopolitics, and Black Feminist Theories of the Human,* Durham, NC: Duke University Press.

2

PureHealth: Feminist New Materialism, Posthuman Auto-Ethnography and Hegemonic Health Assemblages

Tina Sikka

Introduction

I begin this chapter with a brief overview of the state of play vis-à-vis immunity-boosting supplements and COVID-19, followed by an overview and analysis of my generative methodological framework. Note that I ground this study in an understanding of health that is based on my previous work and which I define as 'a co-produced state of idealized expectations, performances, embodiments and patterns of consumption dominated by gendered and raced technophilic knowledge regimes that reproduce regimented and coercive Western standards of health and well-being' or CICT (Sikka, 2022, 2023). Following a précis of this definition, I apply my 'auto-ethnographic and rhizomatic socio-material feminist approach to science and technology' method in four phases, via: (1) a case study-driven overview of Purearth, a UK-based wellness company; (2) a deeper discussion of method; (3) the application of new materialism to the company and its products through the use of an 'agential cut'; and (4) an auto-ethnographic exegesis through which I chronicle and reflect on my consumption of Purearth's 'Immunity Drinks Pack' over approximately ten days.

Case study: Purearth

Purearth is a London-based wellness company (incorporated in 2012) that sells vegan and organic 'water kefirs, cleanses, juices, shots and broths', in recyclable bottles, and whose 'ostensible mission' is to 'help people live healthier lives by using what nature provides us in its most pure and organic form' (Purearth, 2021). The company is funded primarily through Nexus Investment Ventures, a corporate, venture-capitalist seed fund whose latest funding round raised $447,000 for the company (Crunchbase, 2021; PitchBook, 2021). Its owners are two early to middle age White women, Tenna Anette and Angelina, who have been working in the health and wellness industry as practitioners for a number of years. The company has been written up, and its products reviewed, in such outlets as *Country & Town House* (Cox, 2020), which called its cleanse 'one of the most enjoyable restrictive or preset diet plans out there' *Retail Times* (Briggs, 2020), *Wow Beauty*, which invites readers to consume Purearth products and, in doing so, 'drink their way to gut health' (Wow Beauty, 2018) and *The Telegraph* (Howell, 2019), which, in a review, gave its kefir drink a 10/10. Since the COVID-19 lockdowns, Purearth has reported a 254% increase in sales, which is consistent with other data emerging from the health and wellness industry (TechRound, 2020).

I decided to focus on this company and its products for several reasons, beginning, superficially, with the number of times I had seen advertising for it on my social media feeds (likely as a result of doing this research). Serendipitously, it just so happened that the company fit with the ethos of this chapter and reflected dominant health ideologies in ways that brought its non-human-object products, like kefir water, to the fore as 'active' and intelligible (Braidotti, 2006; Fullager, 2017). More *research*-oriented reasons for relying on Purearth include its prominent place within the larger health and wellness ecosystem, the amount of media attention the company has received, its substantial consumer base and its lively social media presence (Twitter, Facebook, Instagram). Finally, Purearth's ethic and brand identity, from its owners to its wellness forward rhetoric, is reflective of analagous companies I also could have chosen including Kroma Wellness, Moon Juice and Flow Water. Purearth was also accessible to me, living and working in the UK, and easy to procure (as I was operating under conditions of constrained movement due to COVID-19).

A final reason for drawing on Purearth for the case study is that its philosophy, mission statement and branding embodies my definition of our normative and dominant conception of 'good health', which, to reiterate, I define as 'a co-produced state of idealized expectations, performances, embodiments, and patterns of consumption dominated by gendered and raced technophilic knowledge regimes that reproduce regimented and coercive Western standards of health and well-being'. Having returned to this definition, I think it is important to take a moment to flesh out this understanding of good health, since, as stated earlier, it provides *the* definitional infrastructure for the remainder of this chapter. CICT sees health as constituted by multiple forms of influence, including (social) media, public health, friends, family, the science/medical community and different layers of government, to produce a model of health that, particularly for women, is oriented towards White, thin and youthful forms of embodiment (note that for men the desirable state tends to be White, youthful, muscular) (White, Young and Gillett, 1995; Stibbe, 2004; Turrini, 2015).

Notably, the enactment of this identity extends far beyond physical embodiment to encapsulate aesthetics, consumption habits and fitness practices. I argue that, in toto, this health regime is: (1) healthist in outlook, wherein personal health is seen as 'a primary – often *the* primary – focus for the definition and achievement of well-being, a goal which is to be attained primarily through the modification of [individualized] lifestyles' (Crawford, 1980, p 366; also see Guthman, 2011; Turrini, 2015); (2) nutritionist, meaning that it reflects a 'reductionist understanding ... of nourishment' that privileges 'universal metrics – calories, nutrients, and so on – redefining "food" (at least for health purposes) to be the sum of these standard parts' (Hayes-Conroy and Hayes-Conroy, 2013, p 1; also see Scrinis, 2008; Biltekoff, 2012); and (3) neoliberal, wherein health is individualized and citizen subjects are made to act as 'rational, calculating creatures whose moral autonomy is measured by their capacity for "self-care" (Brown, 2003; also see McGillivray, 2005; Ayo, 2012).

Meeting these ideological requirements necessitates material enactments that include eating the 'right' foods (often pricey, organic and 'natural'), engaging in the newest health practices (for example, fitness classes, diets, cleanses) and using the latest technologies (diet and fitness trackers, digital health technologies – the technophilic component of CICT). It is important to note that this hegemonic understanding of health is also intensely ableist, western-centric, heteronormative and disciplining – especially for women. Health assemblages, while constantly in flux, can become ossified when forces, flows, objects and ideas become aligned towards specific ends and are made to conform to particular interests. These forces and interests, as Petersen and Lupton attest, are bound by 'a complex and expanding apparatus of control, discipline and regulation that involves micropolitical processes' and pressures to conform, as well as economic motivations that profit from the sale of what are superfluous but lucrative health products and services (Petersen and Lupton, 1996, p 14).

While I engage in a more detailed examination of the ways in which Purearth materially and discursively enacts this dominant definition of good health in the next section, it is important to underline that this health ideology (itself being a contingent outcome of institutional interests, forces and flows) has become concretized in ways that sustain socio-material relations of power. Even a cursory scan of Purearth's website reflects this, from its naturalist aesthetics to its emphasis on the 'wholeness', 'purity' and 'freshness' of its products. Each of these speaks directly to CICT's 'idealised expectations, embodiments, and patterns of consumption'.

Additionally, Purearth's integration of formalized expertise and product-based sociotechnical rigour (albeit with some contradictions), is illustrative of a base technophilia defined through the lens of technological determinism in which technology and progress are seen as mutually constituted (Marx, 1994; Didziokaite, 2017). Finally, throughout the website, and in some promotional material, Purearth's use of images of largely young, White, well-off and normatively attractive women, as well as its choices around languaging, promote a specific kind of health subjectivity, namely, one that is fit, functional, disciplined, normatively attractive and consistent with Western knowledge regimes (Griffith, 2012).

A note on method

FNM, one of the two methods used in this chapter, is an approach to the study of the social world that bridges the chasm between discourse and matter, as well as the body and culture, in ways that seek to recalibrate the relation between ontology and epistemology. Here, 'knowing is [seen as] a [collective, processual and] distributed practice that includes the larger material arrangement' (Barad, 2007, p 379) and where there is 'no "I" separate from the intra-active becoming of the world' (Barad, 2007, p 394). In addition to Karen Barad (2004, 2007), scholars like Rosi Braidotti (2011, 2019), Donna Haraway (2004), Jane Bennett (2010a), Elizabeth Grosz (2005), and Iris van der Tuin (2008) have developed this formative 'onto-epistemology' by integrating the materiality of bodies, the agencies of non-human nature, global entanglements and changing relations of power. FNM builds on poststructuralist and posthuman epistemologies in order to transform existing ontological categories into ones that seek to radically rethink nature and human-nature intra-actions. New materialism seeks a 'return' to matter – thereby 'reflecting a material reality, a theoretical field that goes beyond an anthropocentric view of the world' (Casselot, 2016, pp 77–78), where the subject is seen not as rational, bounded and autonomous, but as 'an amalgam, a collection of heterogeneous components, a material informational entity whose boundaries undergo continuous construction and reconstruction' (Hayles, 1993, p 3).

The 'feminist' element of FNM is reflected in the way in which feminism, and with it new materialism, addresses issues around power, relationality, the dissolution of binaries, inequality, (gendered) bodies, nature, the non-normative Other and the marginalized (Bennett et al, 2010; Lupton, 2019a). Moreover, many of the aforementioned scholars self-identify as feminists and have been working for decades to find ways of articulating how gender, race, class, sexuality and dis/ability intra-act transversally to shape knowledge production 'by activating its ethico-politics', to 'diagnose, infer, and transform gendered, environmental, anthropocentric, and social injustices from a multidimensional angle'. 'Social injustices', Revelles and Rogowska-Stangret write, 'are a driving motivation to pursue research, and are the ... reason [why one] cannot understand new materialism without feminism' (Revelles and Rogowska-Stangret, 2019, p 296).

I use FNM in order to enact 'cuts' that, for a moment, stabilize this complex health assemblage and, in doing so, operate 'as a material-discursive boundary-making practice that produces "objects" and "subjects", and other differences out of, and in terms of, a changing relationality'(Barad, 2007, pp 92–93). This approach fosters a unique way in which to think about how we are entangled with technologies (inclusive of the substances we consume), and brings to the fore phenomena that have been neglected by applying a generative mix of socio-discursive and material analysis. Auto-ethnography delivers a complementary layer of critique by drawing on reflexive *and relational* introspection, reinforced by years of training and scholarship, which I then filtered through inequality, injustice and power relations. My approach to auto-ethnography is firmly in line with other 'layered methods' that involve data collection (through copious note taking) and the use of literature and personal narrative (Ellis, Adams and Bochner et al, 2011; Reed-Danahay, 2021). It is these two components that co-produce the 'auto-ethnographic and rhizomatic socio-material feminist approach to science and technology' method that I have become committed to deploying in my research.

Feminist new materialism: agential cutting

As noted, agential cuts are contingent boundary-making practices that account for intra-actions in assemblages that mark bodies. They are also knowledge-producing practices that materialize different aspects of a phenomenon for which the researcher is responsible. As Barad states, 'the world can never characterize itself in its entirety; it is only through different enactments of agential cuts, different differences, that it can come to know different aspects of "itself"' (Barad, 2007, p 432; also see Barad, 1998). The agential cut I have made and will focus on is as follows.

PureHealth − body normativity − immunity − neoliberalism

There are several ways in which cuts can be made, since all perform intra-actions that are powerful. Most demand less of a disentangling, but, instead, the careful 'magnifying [of] details through a lens' (Kautz and Jensen, 2013, p 94). The boundary-making tools I have applied are analytic, discursive and embodied and, like any other research instrument, are knowledge producing. Which is to say that cuts enact and identify embedded relations of power while also bringing to the fore the vibrancy and agency of non-human artefacts (Bennett, 2004; 2010b). Normative embodiment, immunity and neoliberalism are diffractively related to one another and invite the interrogation of CICT as it emerges out of Purearth both as a company and as a brand. This cut reflects, in its own way, the deep and situated nature of 'the movements between meanings and matter, word and world, interrogating and re-defining boundaries … in "the between" where knowledge and being meet' (Barad, 1996, p 185; also see Fox and Alldred, 2016).

The cut

PureHealth – body normativity – immunity – neoliberalism

This cut provides an opportunity to elaborate on PureHealth's focus on specific kinds of bodies and the relationship between bodies and health. To begin with, the normative bodies that PureHealth cultivates are gendered as female (again, this company's target demographic is women), White, youthful, thin, and demand the disciplined adherence to both 'nutritionist' and 'healthist' discourses and embodiments (Selk, 2002; Azzarito, 2012). In this case, however, CICT health discourse, when coupled with the direct hailing of complex subjectivities under the rubric of responsibilized health (and expectations around acceptable bodily comportment), reflects *material-discursive* forces and flows that co-construct the desirable female body. PureHealth, on its website, makes liberal use of this kind of languaging when it speaks about the desirable body. This includes its promise that their cleanse will 're-set' and 'heal' your body, leaving you 'feeling revitalised and lighter', as well as their stated commitment to help customers 'live healthier lives' 'through practices of detoxification, hydration, and replenishment' (Purearth, 2021). What is notable here is that discourse that might have been more direct in the past, for example, by promising weight loss and a thin, desirable and attractive body, is now less inclined to use such overt rhetoric. This reflects a very real socio-material change in norms around what is permissible with respect to health discourse in which wellness, well-being, healing and balance are used to mask the persistent policing of body size and the saliency of diet culture (Bordo, 2004; Bartky, 2015). While seemingly freely 'chosen', these practices remain committed to cultivating docile bodies through ascetic and self-disciplining directives (Heyes, 2006; Rose, 2009; Germov and Williams, 2017).

Moreover, Purearth's underlying ethos is 'healthist' and 'nutritionist', which together work to enact and solidify CICT and the normative Western body. Healthism is a term coined by Robert Crawford to refer to a 'new health consciousness' in which personal health takes on an increasingly influential role, becoming 'not only a preoccupation; it has also become a pan-value or standard' (Crawford, 1980, p 67). More contemporary applications of healthism draw attention to how health has become auditable, hyperdisciplined and imperative for the active, responsiblized, labouring citizen-subject (Cheek, 2008; Lupton, 2013). Healthist health is also embodied, biosocial and reflective of hierarchies of power and privilege related to class, gender and race. Finally, according to this logic, health has come to reflect a supervalue and a measure one's moral worth wherein the barometer of success is *materially inscribed on the body* (Guthman and DuPuis, 2006; Cairns and Johnston, 2015; Rodney, 2018). PureHealth's emphasis on how their products will 'have a powerful effect on your health, mood and ultimately your quality of life' through the consumption of their 'immune boosting, gut loving, life changing drinks + more' (Purearth, 2021), must be understood as standing in relation to CICT as a healthist assemblage in which what we eat and how we move has taken on a moral valence. While the discourse might have changed, healthism continues to demand a technology of self that results in similarly hegemonic health subjectivities and embodiments (that is, the normative body).

The other component of body normativity that makes itself felt is in relation to an underlying nutritionism in which particular nutrients are made to take on a magical quality by promising an outsized effect on health (Kimura 2013). Fortified foods, superfoods and other therapeutic ingredients, for example, have been used to encourage a quantified dietary biopolitics wherein 'eating right' produces desired health outcomes (Scrinis, 2008; Biltekoff et al, 2014; Loyer, 2017). With respect to Purearth, those magical ingredients are organic, high in 'micro and macro nutrients, providing your body with a higher quantity and quality of minerals, vitamins, enzymes and nutrients'. Grygory Scrinis warns against this myopic focus on individual nutrients since they are often 'abstracted out of the context of, and analyzed in isolation from, other nutrients and food components, as well as from foods and dietary patterns' (Scrinis, 2012, pp 271–272). Purearth sells this logic through their 'cold-pressed' fruits and vegetables, which promise to provide 'your body with a higher quantity and quality of minerals, vitamins, enzymes and nutrients', as well as their 'superfood infused shots' which, it is claimed, will 'reduce tiredness and support overall physical and mental wellbeing' (PureHealth, 2021).

Hey writes about a particular form of contemporary nutritionism that is especially applicable here called 'healthist fermentation', which refers to the 'myopic view of ferments as a functional food or tool, capable of delivering specific nutrients or metabolites, including their preparation and consumption, for the explicit purposes of supplementing one's physiological makeup' (Hey, 2020, p 14). This can be seen in Purearth's water kefirs which are (Turkish) fermented drinks that promise to replenish gut bacteria: 'Containing over 27 billion live cultures per bottle, including live strains of Lactobacillus and Bifidobacterium (probiotics) our Water Kefir is teeming with beneficial digestive enzymes, amino acids, vitamins, and minerals that our bodies love' (Purearth, 2021).

A materialist assessment of the imbrication of food and bodies enacted by this product might look favourably at assertions that highlight how bodies are co-constituted with, by and through the food they consume. As Kass describes it, 'in eating, we do not become the something that we eat; rather the edible gets assimilated to what we are ... the edible object is thoroughly transformed by and re-formed into the eater' (Kass, 1999, pp 25–26). Some of this is evident in how gut health and bacteria are described by PureHealth and yet, as I argue further on, it remains confined to the individual and actuated in pursuit of functional healthist and nutritionist goals.

Together, healthism and nutritionism work to co-construct a socio-material 'healthy' body that, despite the language of wellness and well-being, remains entrenched in the logic of CICT-based health norms. This, as the images of slim, young, attractive women (and one man) on Purearth's website attest, has simply replaced diets with lifestyle change, thereby essentially encouraging a 'forever diet'. Their objective, however, remains the same, wherein 'wellness seekers engage in a profoundly moral discourse around health promotion, constructing a moral world of goods, bads and should' where the body is a 'site for moral action' (Conrad, 1994, p 385). Noncompliance is thus seen as a personal failure and often results in self-punishment in the form of stringent diets and over-exercise as well as fat-stigma (Calogero, Tylka and Mensinger, 2016; Pausé, 2017).

Immunity, the next element of this cut, has its own set of socio-historical, material and discursive forces and flows that reflect apprehensions around risk, self-control, protection from Others (both viruses and people), bodily integrity and sanitation (Crawford, 2006; Metzl, Kirkland and Kirkland, 2010; Bliss, 2015; Haraway, 2020). The marked rise in spending on so-called 'immunity-boosting' supplements since 2020 reflects concerns enacted by COVID-19 regarding health and illness, with supplements sold as 'natural' claiming to provide an alternative or complement to vaccines. Purearth, like many of its counterparts, never explicitly mentions COVID-19 or any other disease or infection in this context, which would risk running afoul of data laws, but, instead, does so indirectly by construcing immunity as a 'consumable' product. Purearth, for instance, claims that its shots (one of which is called 'Immunity Boosting Hot Shot') can be taken when you 'feel like you need that little extra boost' or are 'feeling cold' – implying that immunity is discrete property. Elsewhere, Purearth labels one of its cleanses the 'Immunity Reboot Juice Cleanse', which promises to 'flood your body with lush green juice, essential vitamins and minerals, pre + probiotics, and medicinal properties to support a healthy immune system and overall vitality' (Purearth, 2021). While much of this languaging existed prior to COVID-19, I argue that it reflects a larger socio-material and discursive trend in which a sleight of hand is used to activate the embodied anxieties of the 'worried well'. In a more recent addition to the website, Purearth has inserted a note about the introduction of cannabidiol (CBD) shots and the ramping up of the production, in 2020, of 'our immune boosting drinks in the midst of a global pandemic'. In doing so, they avoid any explicit assertion of a connection between the two (that is, their immunity-boosting products and COVID-19), but imply that link nonetheless.

This framing aligns with suspicions around Western biomedicine including the not misplaced belief that it has been and can be exploitative, patriarchal and fail to 'serve women's health needs, which are often under-researched … and undertreated … while women's bodies are over-medicalized' (Liedy Sievert, 2003; Mikhail, 2005; Shahvisi, 2019, p 99). For women this differential treatment, ranging from dismissal to disbelief, has helped to produce attenuated CICT subcultures in which holistic, alternative, healthist and consumption-based health practices are seen as viable alternatives to modern medicine (Greenwood, Carnahan and Huang, 2018; Davis, 2019; Crockford, 2020). Moreover, understandings of biomedicine as iatronic and bureaucratic have produced health ideologies that promise material social change but end up promoting individualized conceptions of health rooted in consumerism and status seeking (Baer, 2004). Immunity thus becomes a focal point of these communities with, for example, higher-income, predominantly White, young mothers opting to forgo vaccines entirely (Winter, 2020; Phelan, 2021).

What is unique about how vaccine discourse has evolved at this particular juncture is that it reveals the deep entanglements between illness and the body wherein the latter is experienced as bounded and contiguous and vaccines are seen as foreign and violative (Sanders and Burnett, 2019). This throws a monkey wrench into the technophilic dimension of CICT, since scepticism of the vaccine is not particularly technophilic. However, I would maintain that this faith has merely been displaced from the hegemonic to the alternative wherein health is cultivated, enacted and worked on 'through heath protective behaviours' that seek to mitigate a 'semi-pathological pre-illness at-risk state' (Armstrong, 1995, p 401). Consequently, 'expertise' becomes linked to experiential knowledge – which can be discursively compelling and is reflective of feminist methodologies. Purearth draws on this ethos through the owner's personal back stories and the company's ethos of passion and care. Vaccines, in this environment, tend to be seen as sterile, asocial and imposing. Compare that with images of Purearth's owners 'spending early mornings and late evenings washing, peeling and juicing, then testing all our products on family and friends to ensure we had the tastiest, healthiest drinks' (Purearth, 2021).

What we have in this part of the cut is a study of immunity which takes on and adjusts CICT for specific ends while also attending to how health, as an embodied and resonant state of relational being, can be complexified through worries about contagion, Otherness, bodily integrity, expertise, forms of knowledge production and trust. Purearth, through its use of a discourse of care and confidence, stitches together a narrative that is compelling – particularly its promise of a bodily, material experience of well-being through pleasurable embodied encounters with its lively and vibrant products (Bennett, 2010a and b; Meskus and Oikkonen, 2020). Vaccines, conversely, fall short and fail to attend to an understanding of our body-selves as lived, productive and complex (Deleuze and Guattari, 1988; Lupton, 2019b).

The final component of this functional cut is neoliberalism, which takes on a particular form of saliency as it relates to Purearth by operating as a constitutive material-discursive practice that functions in a similar way to measuring devices (Barad, 2012). CICT is deeply neoliberal, wherein neoliberalism is used to characterize phenomena including: (1) the extension of economic logic to the social, cultural and public spheres; (2) moves towards increased privatization, deregulation, marketization and commodification (accompanied by cuts to social programmes); and (3) the concomitant construction of the individual as an autonomic, rational, self-interested economic actor (Harvey, 2007; Chomsky 1999; Giroux 2005; Duggan 2012).

With respect to health, the influence of neoliberalism has produced health subjects that are made responsible for their own health status through a process of intense responsibilization as a result of sheer necessity in the light of the dismantling of social safety nets. Alkon (2014) describes the neoliberal health subject as required to enact food and exercise behaviours that turn the individual into a moral entrepreneur, ideal citizen subject, and tool of neoliberal governmentality wherein choice reigns supreme (Ayo 2012; Cairns and Johnston, 2015; Lupton 2017). This neoliberal ethos encourages self-surveillance (that is, technologies of self) and biopedagogies requiring action on the material body so as to reproduce subjects with an ideal level of health capital (Foucault, Davidson and Burchell, 2008; Foucault, 1988; Wright and Harwood, 2012). Purearth's entire ethos is about helping customers make those 'correct' choices by allowing them to shop by health concern and curate their own health packs, and by offering a consumer-based way to make changes in their 'lifestyle choices'. This includes by drinking juices that purport to cleanse 'the body by removing toxins and heavy metals from the body' and with the promise that customers will

feel fully tuned into your body, eliminating all forms of digestive discomfort and helping reduce inflammation. Your mind will feel clear and collected, you may even cleanse on an emotional level releasing pent up negative energy. Expect healthy weight loss as your body decomposes cells in excess and renews on a cellular level. Cleansing is a great way to kick-start a healthy eating regime, feeling fuelled with long-lasting energy! (Purearth, 2021)

This framing of neoliberal choice as *the* way in which to become 'healthy' is also in keeping with larger social goals around entrepreneurism, the production of ideal labouring subjects, and values including responsibility and hard work. Packaging health in the language of empowerment and autonomous choice is also postfeminist in that management of the body is a moral imperative – one that 'instructs and regulates young women's bodies and subjectivities through a language of choice, empowerment and health while ... exercise [in this case diet] is conceived as a discipline to achieve the normative body' (Camacho-Miñano and Gray, 2021, p 726; also see McGregor, 2001; Gill and Scharff, 2013; Riley, Evans and Robson, 2018). This is particularly in keeping with CICT health ideology and the ways in which Purearth, neoliberalism and CICT function to set precise expectations around embodiment, affect and well-being.

It is also the case that neoliberal health formations have material, bodily effects including anxiety and disconnection from others. Conversely, for those with means (likely most of Purearth's customers), the promise of 'embodied contentment' that permeates Purearth and larger wellness culture can feel empowering. Specifically, it is the promise of cultivating connections that are vibrant by engaging with substances that have the power to 'boost energy, alkalise, beautify' and 'nourish, detox and strengthen' that is compelling (Purearth, 2021). The body, while retaining its neoliberal individualism, is then taken up as corporeal, as a living enterprise that resonates with the organic, natural and health-sustaining Purearth products it consumes (Bennett, 2010a and b; Coole and Frost, 2010).

This is consistent with a feminist-materialist ethic but lacks a sense of material connection with the wider natural world – what Massey (2005) refers to as the 'relational constructiveness of things' and, for Lawson, the variegated 'connections that bind us together' (Lawson, 2007, p 4). Dolphijin and van der Tuin (2010) refer to this as the transversal relation in which the self is produced in material-semiotic networks that are relational all the way down. Purearth limits this relationality by solidifying a neoliberal, CICT health subjectivity in which a more capacious understanding of health are delimited.

This cut has brought to the fore a number of salient themes and issues around body normativity, immunity and neoliberalism, including that of the disciplined body, diet culture, healthism, nutritionism, risk and health anxiety, scepticism around Western biomedicine, wellness culture, individual responsibilization and self-surveillance. As such, it has produced an 'enacted cut' – a considered (re)configuring of shifting boundaries that form around material-discursive practices and which, in this case, involve the analysis of a particular dimension of the CICT–Purearth health assemblage that is diffractive and assumes the inseparability of 'subject and object, nature and culture, fact and value, human and non-human, organic and inorganic, epistemology and ontology, and material and discursive' (Barad, 2007, p 333). In the next section, I use a post-subjective or posthuman form of auto-ethnography to probe aspects of this health assemblage further. This approach is embodied and affective and one in which I, as the researcher, use myself 'as an object of description, analysis, and/or interpretation' (Chang, 2013, p 35), and where experience serves as data.

Posthuman auto-ethnography: Purearth and CICT

As noted earlier, the kind of auto-ethnography I employ draws on reflective analysis in the form of diary entries, note taking and voice notes to make meaning out my experience of consuming PureHealth's 'Immunity Drinks Pack' (which comes with immunity juices, water kefirs, immunity booster shots and vegan broths). Engaging with these products acts as a useful way to connect the personal to the social and, in doing so, unsettle aspects of CICT in ways that make room for alternatives. It is important to underline that the form of auto-ethnography I am drawing on is posthuman and anti-solipsistic, wherein the 'subject' is seen as intersectional, co-constituted and decentred. As Alsop asserts, subjectivity, understood in this way, emerges out of the 'crisscross[ing] between the boundaries ... of being insider and outsider [and] of being personal and cultural selves' (Alsop, 2002, p 13). Kate Warfield takes this one step further by arguing for auto-ethnographies in which 'The first-person voice of the self may become multiple, fragmented, or incomplete and in motion – that is, uncertain', and where power, activism and jarring are par for the course (Warfield, 2019, p 153). Objects and substances, as we will see, 'become the thing that challenges us to explain the assemblages of life' (Dickinson, 2017, p 86). In the following paragraphs, I provide a brief description of the package purchased before taking up the central themes, provocations and challenges that came up in the course of what was approximately ten days of consuming these products. My analysis and reflections centre on what consuming them evoked in me, rather than on how they taste or whether they 'worked'. I specifically take up and discuss cultural appropriation, gender and race (and their intersections).

As per my routine, I committed to drinking at least one, preferably two, items in the pack per day, with the juices first – due to their limited shelf life. I consumed the juices early in the day, with the water kefirs and shots later in the afternoon. The broths I found worked well as a base for noodles, which I ate in the evening. In terms of enjoyment and taste, all of the items were pleasant to enjoyable, with the sparkling water kefirs tasting particularly good.

Upon receiving the pack and opening it, the first thing I found especially interesting was Purearth's description of their 'Immunity Drinks Pack', which stated that it provides 'support [for] your immune system' by 'nourish[ing] your gut microbiome'. Being suspicious by nature, I tried to find out when this option was introduced and whether it aligned with COVID-19 (unfortunately without success). In reading up about the included products, what struck me next was their appeal vis-à-vis one of the included products which mentioned the healing properties of Ayurvedic medicine based on their use of turmeric and ginger, which, they assert, has 'been used ... for centuries to help treat inflammation, detoxify and regenerate the digestive system' (Purearth, 2021). This brings me to one the main themes I returned to repeatedly in my analysis – namely, authenticity, cultural specificity and cultural appropriation.

Purearth's liberal use of terms like 'Ayurveda', 'Thai spice', and kefir to describe their products felt to me as in keeping with critiques of cultural appropriation that have emerged since 2020. The problem I have with this has less to do with the consumption of foods that are not autochthonous to one's traditional foodways (which is absolutely fine) and more to do with *how* this is done. Ayurveda, for example, is a thousand-years-old, complex applied philosophy and socio-religious practice still used today in South Asian households throughout the world. It also has its own regional and institutional specificities, with varying claims to scientific authority, medical rigour and disease treatment and prevention (Alter, 1999; Khalkova, 2018). I myself, being of South Asian descent, remember using face masks and consuming warm drinks made with turmeric and ginger as well as listening to the claims made by my grandmother about the curative properties of these foods. The cultural specificity, history and context of these practices, when taken and inserted into a new terrain for the purpose of selling products – as in the case of PureHealth, can be appropriative and neocolonial in that it permits: (1) the safe consumption of Otherness; and (2) a de-politicized experience of Otherness that delimits risk (hooks, 2012; Julier, 2016; Gertner, 2019).

Additionally, both ginger and turmeric, when consumed through these drinks, can be seen as a means through which dominant, White communities can perform cosmopolitanism and secure their liberal bona fides while retaining the knowledge that, in the end, they are always able to return to Whiteness. The histories of colonial subjugation, profit, exploitation and the past casting of certain food and ingredients as 'gross', 'smelly' and 'exotic' (but which are now are sold at a mark-up) felt, to me, as intentionally elided. As such, I felt annoyed when consuming the Purearth products that appropriated the discursive and material richness of non-indigenous cultural practices to turn a profit (Johnson, 2003; Nguyen and Strohl, 2019).

This experience led me to re-read a particularly salient article I had come across some time ago in *Overachiever Magazine* titled 'The Cultural Appropriation of Turmeric', in which the author traces the use of turmeric from its historical roots to Western researchers who were increasingly 'poaching on ... ancient healing techniques' and 'misappropriating generations-old traditional knowledge'. Also relayed is the transformation of turmeric from *the* 'spice behind all of this negativity, cultural stereotyping, and bullying' to one that 'has recently risen to sudden glamour and fame' (Joshi, 2020). The beneficiaries of this appropriation, like the owners of Purearth, are largely White, affluent, able bodied, cis-gender and straight-sized women. Moreover, this capital, accrued through what Grey and Lenore (2018) call 'culinary colonialism', is part of the capitalist and neoliberal 'marketing and selling of diversity' which has become de rigeur (Abu-Laban and Gabriel, 2002). A similar kind of analysis can be applied to Purearth's sparking water kefir, which is a drink I quite enjoyed (it tasted refreshing, particularly on a hot day). Kefir, however, like turmeric, has its own deep and rich history rooted in the North Caucasus going back centuries, when nomads would mix kefir grains and milk inside goatskin bags to drink on their journeys (Farnworth, 1999). While I am not able to examine this further here, what is clear is that kefir, like turmeric and ginger, has been decontextualized (even having the dairy removed from it) and inserted into a CICT-centric assemblage that works to make the 'exotic familiar' and interpose Purearth as the cultural intermediary that shares this insider knowledge (Heldke, 2001).

Another theme that came up several times was around race, gender and their intersections (Nash, 2008; Crenshaw, 2017). In particular, my notes reflected on how Purearth's products enacted racializing logics and gendered embodiments. First, I found it interesting that the gendered subjectivity they cultivated in me felt very much tied to the coding of fasts, juicing and food restriction as feminine acts. Femininity, dieting and the policing of bodies is historically rooted and continues to be reflected in advertising and media (Ling 2015; Lau, 2016), as well as in broader social attitudes wherein guilt, purification and redemption are tied up with eating behaviours (Spoel, Harris and Henwood, 2012). I began to think about how dieting became an early part of my relationship to my body in which what it meant to be attractive and desirable was reflected in performances of constrained eating (but carefully orchestrated so as to not appear too restrictive). Johnson and Cairns refer to this as the 'do-diet', which they define as 'a practice in which women actively manage their relationship to the extremes of self-control and consumer indulgence in an effort to perform acceptable middle-class femininities' (Cairns and Johnston, 2015, p 156). The cleanses Purearth sells sit at the far end of this spectrum, since they would require the avoidance of any social situations with food, but the addition of specific drinks products might not necessitate an intense amount of adjustment (although I did note that it had an effect on my other food choices which began to fall into line with Purearth's 'clean', 'natural' and 'unprocessed' ethos).

Some of my earlier entries reflected on the fact that pressures around femininity did not feel causal or self-organizing – that is, the products did not feel *feminizing* in and of themselves. Rather, I observed that perhaps the best way to think about them was in relation to the ways in which they worked with other institutions, technologies and cultural pressures to maintain the dominance of the slim, White, disciplined body as *the* feminine ideal (Bartky, 2002; Bordo, 2018; Lupton, 2018). Very quickly, however, it became clear that what these drinks promised and promoted was actually much more powerful. By endorsing self-discipline as empowerment, PureHealth drew on the postfeminist and neoliberal ethic discussed earlier quite deftly, wherein restriction becomes liberation, choice functions as the ultimate indicator of progress and 'transformation is constructed as empowering by offering self-mastery, health, a moral position and associated citizenship, and hope/ expectations of a good life' (Riley and Evans, 2018, p 213). It is important to underline that this form of femininity is, by default, always able-bodied and heteronormative.

With respect to race and racialization, I was very much cognizant of the Whiteness of the space PureHealth cultivated online in light of their choice of images, models and aesthetics. However, I was surprised by how fully this permeated into the notional wellness space on the level of 'common sense' wherein Whiteness felt 'constructed as universal and unmarked, not particular or exclusive' (Dosekun, 2020, p 101; also see Dyer, 1997). I noted that CICT, as it is expressed in this space, functions as a racializing technology – one that upholds and reinscribes racial hierarchies. Bodies that are ideal, acceptable and healthy reflected a kind of Whiteness that, it became clear, I did not and could not have (Collins, 1990; Delgado and Stefancic, 2013). As a straight-sized woman of South Asian descent, I felt quite alienated from this space. I wrote about how this made me feel with respect to what had been constructed as 'normal' – namely the image of the end user who has acquired a habitus, capacities, priorities and set of basic characteristics – ones that I did not and could not possess. The monetary capital and cultural knowledge required to participate in these trends also cleaves along racial lines and follows previous exclusions of racialized women from particular networks and practices (Cole and Sabik, 2009). The work of scholars like Sabrina Strings (2019) and others has drawn attention to this process of 'whitewalling' (D'Souza, 2018) or 'white racial framing' (Feagin, 2013), wherein fatness intersects with Blackness and material-discursive practices to maintain White-only spaces. Despite my level of education, cultural capital and class position, race continued to function as a technology of exclusion. Towards the end of the ten days, I reflected on how race and gender overlapped in notable ways – particularly as it relates to how they work synergistically to construct the body as something to be maintained and disciplined. Purearth and its enactment of CICT, I wrote, 'sort of makes me feel pulled in two directions – i.e. forcibly interpellated as a woman and then alienated as a woman of colour'.

Finally, something I was most surprised by in putting together this auto-ethnography, was how secondary the experience of consumption was in my own writing. FNM, in its focus on embodiment and physicality (where both are experienced as relational phenomena), should elicit insights around sensory experience, affective threads and the mediation between body and substance (Massumi, 2002). However, my analysis and observations tended to be tied less to the acts and experiences that surrounded consuming the kefir drinks, which I enjoyed the most, the juices and shots, which were pleasant, and the broths, which were also quite good, and, instead, focused on the sociopolitical. This is itself an important insight, one that speaks to the tendency to discount material embodiment and engagement – what Sara Ahmed (2013) calls 'sites of contact'. One of the most resonant applications of this approach, lacking in my own analysis, comes from Jane Bennett, who writes persuasively about the agency of edible matter wherein 'food [operates] as a co-participant in our world, as possessing an agentic capacity irreducible to (though rarely divorced from) human agency' (Bennett, 2007).

For one reason or another, the posthuman element of this auto-ethnography became subsumed, for me, by a focus on representation. The one area in which a form of embodied relationality did coalesce was in relation to the water kefir drink about which I observed, in a voice memo, how odd it was to think about the process of fermentation and the idea that this substance would be feeding me and my microbiome in perpetuity. This, I wrote, speaks to the fact that our bodies are both human and non-human at the same time. Or, as Bennett eloquently puts it, the imbrication of bodies and others means that 'My flesh is populated and constituted by different swarms of foreigners ... we are, rather, an array of bodies, many different kinds of them in a nested set of microbiomes' (Bennett 2010b, pp 112–113).

This auto-ethnography has produced several insights around culture, gender, race and the experience of embodiment grounded in my affective and relational experience of Purearth's products as well as its grounding in CICT as the dominant health ideology. Issues around Whiteness, femininity, bodily regulation and the mediated body have been discussed in order to bring to the fore the importance of narrative as a way to take up the social and material through the personal such that 'bodily ... unbeknown knowledge' is unearthed via auto-ethnographic writing in order to reveal 'those practices that have become invisible because of their ordinariness or repetitiveness' (Uotinen, 2011, p 1309).

Conclusion

Drawing on Purearth as a case study, this chapter has used FNM and posthuman auto-ethnography (auto-ethnographic and rhizomatic socio-material feminist approach to science and technology) to unpack how CICT, as our dominant health ideology, functions to produce a conception of health that is exclusionary, disciplining, racialized and gendered. I began by examining how our dominant health ideology (CICT) functions in practice using Purearth as a case study, first by applying an agential cut to study particular aspects of this assemblage. I then followed this with an auto-ethnography in order explore my own emergent experience with Purearth's products. In using this hybrid approach, I have been able to unpack how: (1) a coercive form of body normativity operates through the rhetoric of healthism and nutritionism; (2) how immunity functions as a site of risk, worry, scepticism and foreignness; and (3) how neoliberalism works to responsibilize health subjects by instilling an ethic of self-surveillance and the illusion of autonomous choice.

In my auto-ethnographic analysis, I examined and discussed such themes as cultural appropriation, the ways in which Purearth enacts a particular form of gendered femininity, and how it and CICT are both racialized and racializing. While limited in its analysis (much more could be said about class, sexuality, dis/ability and their socio-material intersections), this chapter has attended to and explored important sites of disjuncture, inconsistency and potential rupture around what it means to be in 'good health'. This, I argue, can function as a source of opportunity for rethinking health and challenging the hidden dimensions of inequality that continue to be reproduced. The work being done to promote fat liberation/justice (McAllister, 2009; Friedman, Rice and Rinaldi, 2019), Black veganism (Harper, 2009; Wrenn, 2019) and health justice (Fox et al, 2016; Hobart and Kneese, 2020), are just three of several movements that we should look to, going forward.

References

Abu-Laban, Y. and C. Gabriel (2002) *Selling Diversity: Immigration, Multiculturalism, Employment Equity, and Globalization*, Peterborough: Broadview Press.

Ahmed, S. (2013) *The Cultural Politics of Emotion*, New York: Routledge.

Alkon, A.H. (2014) 'Food justice and the challenge to neoliberalism', *Gastronomica: The Journal of Food and Culture*, 14(2): 27–40.

Alsop, C.K. (2005) 'Home and away: self-reflexive auto/ethnography', in W.M. Roth (ed) *Auto/biography and Auto/ethnography*, Leiden: Brill Sense, pp 403–420.

Alter, J.S. (1999) 'Heaps of health, metaphysical fitness: Ayurveda and the ontology of good health in medical anthropology', *Current Anthropology*, 40(S1): S43–S66.

Armstrong, D. (1995) 'The rise of surveillance medicine', *Sociology of Health & Illness*, 17(3): 393–404.

Ayo, N. (2012) 'Understanding health promotion in a neoliberal climate and the making of health conscious citizens', *Critical Public Health*, 22(1): 99–105.

Azzarito, L. (2012) 'The rise of the corporate curriculum: fatness, fitness, and whiteness', in J. Wright and V. Harwood (eds) *Biopolitics and the 'Obesity Epidemic': Governing Bodies (Vol. 3)*, London: Routledge, pp 91–204.

Baer, H.A. (2004) *Toward an Integrative Medicine: Merging Alternative Therapies with Biomedicine*, Lanham: Rowman Altamira.

Barad, K. (1998) 'Getting real: technoscientific practices and the materialization of reality', *Differences: A Journal of Feminist Cultural Studies*, 10(2): 87–91.

Barad, K. (2007) *Meeting the Universe Halfway: Quantum Physics and the Entanglement of Matter and Meaning*, Durham, NC: Duke University Press.

Barad, K. (2012) 'On touching: the inhuman that therefore I am', *Differences*, 23(3): 206–223.

Bartky, S.L. (2002) 'Suffering to be beautiful', in K.P. Addelson, S. Bordo and S.L. Bartky (eds) *Gender Struggles: Practical Approaches to Contemporary Feminism*, Lanham: Rowman & Littlefield, pp 241–256.

Bartky, S.L. (2015) *Femininity and Domination: Studies in the Phenomenology of Oppression*, London: Routledge.

Benavente, B.R., Rogowska-Stangret, M. and Ernst, W. (eds) (2020) *Feminist New Materialisms: Activating Ethico-Politics through Genealogies in Social Sciences*, Basel: MDPI.

Bennett, J. (2004) 'The force of things: steps toward an ecology of matter', *Political Theory*, 32(3): 347–372.

Bennett, J. (2007) 'Edible matter' *New Left Review*, 44, https://newleftrev iew.org/issues/ii45/articles/jane-bennett-edible-matter.

Bennett, J. (2010a) 'A vitalist stopover on the way to a new materialism', in J. Bennett, P. Cheah, M.A. Orlie and E. Grosz, *New Materialisms: Ontology, Agency, and Politics*, Durham, NC: Duke University Press, pp 47–69.

Bennett, J. (2010b) *Vibrant Matter*, Durham, NC: Duke University Press.

Bennett, J., Cheah, P., Orlie, M.A. and Grosz, E. (2010) *New Materialisms: Ontology, Agency, and Politics*, Durham, NC: Duke University Press.

Biltekoff, C. (2012) 'Critical nutrition studies', in J. Plicher (ed) *The Oxford Handbook of Food History*, Oxford: Oxford University Press, pp 172–190.

Biltekoff, C., Mudry, J., Kimura, A.H., Landecker, H. and Guthman, J. (2014) 'Interrogating moral and quantification discourses in nutritional knowledge', *Gastronomica: The Journal of Food and Culture*, 14(3): 17–26.

Biss, E. (2015) *On Immunity: An Inoculation*, Melbourne: Text Publishing.

Bordo, S. (2004) *Unbearable Weight: Feminism, Western Culture, and the Body*, Berkeley, CA: University of California Press.

Bordo, S. (2018) *Reading the Slender Body*, New York: Routledge.

Braidotti, R. (2006) 'Posthuman, all too human: towards a new process ontology', *Theory, Culture & Society*, 23(7–8): 197–208.

Braidotti, R. (2011) *Nomadic Theory: The Portable Rosi Braidotti*, New York: Columbia University Press.

Braidotti, R. (2019) *Posthuman Knowledge*, Cambridge: Polity Press.

Briggs, F. (2020) 'Purearth sets out the top food and drink trends for 2021', *Retail Times*, 14 December, www.retailtimes.co.uk/purearth-sets-out-the-top-food-and-drink-trends-for-2021/.

Brown, W. (2003) 'Neo-liberalism and the end of liberal democracy', *Theory & Event*, 7(1), https://doi.org10.1353/tae.2003.0020.

Cairns, K. and Johnston, J. (2015) 'Choosing health: embodied neoliberalism, postfeminism, and the "do-diet"', *Theory and Society*, 44(2): 153–175.

Calogero, R.M., Tylka, T.L. and Mensinger, J.L. (2016) 'Scientific weightism: a view of mainstream weight stigma research through a feminist lens', in T.A. Roberts, N. Curtin, L.E. Duncan and L.M. Cortina (eds) *Feminist Perspectives on Building a Better Psychological Science of Gender*, Berlin: Springer, pp 9–28.

Camacho-Miñano, M.J. and Gray, S. (2021) 'Pedagogies of perfection in the postfeminist digital age: young women's negotiations of health and fitness on social media', *Journal of Gender Studies*, 30(6): 725–736.

Casselot, M.A. (2016) 'Ecofeminist echoes in new materialism?' *PhaenEx*, 11(1): 73–96.

Chang, H. (2013) 'Individual and collaborative autoethnography as method', in S.H. Jones, T.E. Adams and C. Ellis (eds) *Handbook of Autoethnography*, Walnut Creek, CA: Left Coast Press, pp 107–122.

Cheek, J. (2008) 'Healthism: a new conservatism?' *Qualitative Health Research*, 18(7): 974–982.

Chomsky, N. (1999) *Profit over People: Neoliberalism and Global Order*, New York: Seven Stories Press.

Cole, E.R. and Sabik, N.J. (2009) 'Repairing a broken mirror: intersectional approaches to diverse women's perceptions of beauty and bodies', in M.T. Breger and K. Guidroz (eds) *The Intersectional Approach: Transforming the Academy through Race, Class and Gender,* Chapel Hill, NC: University of North Carolina Press, pp 173–192.

Collins, P.H. (1990) *Black Feminist Thought: Knowledge, Consciousness and the Politics of Empowerment*, Cambridge, MA: Unwin Hyman.

Conrad, P. (1994) 'Wellness as virtue: morality and the pursuit of health', *Culture, Medicine and Psychiatry*, 18(3): 385–401.

Coole, D. and Frost, S. (2010) 'Introducing the new materialisms', in J. Bennett, P. Cheah, M.A. Orlie and E. Grosz (eds) *New Materialisms: Ontology, Agency, and Politics*, Durham, NC: Duke University Press, pp 1–44.

Cox, R. (2020) 'Review: Purearth 3-day gut nourish cleanse', *Country and Town House*, www.countryandtownhouse.co.uk/style/health-and-beauty/review-purearth-3-day-gut-nourish-cleanse/.

Crawford, R. (1980) 'Healthism and the medicalization of everyday life', *International Journal of Health Services*, 10(3): 365–388.

Crawford, R. (2006) 'Health as a meaningful social practice', *Health*, 10(4): 401–420.

Crenshaw, K.W. (2017) *On Intersectionality: Essential Writings*, New York: The New Press.

Crockford, S. (2020) 'What do jade eggs tell us about the category "esotericism"? Spirituality, neoliberalism, secrecy, and commodities', in E. Asprem and J. Strube (eds) *New Approaches to the Study of Esotericism*, Leiden: Brill, pp 201–216.

Crunchbase (2021) Purearth, www.crunchbase.com/organization/purearth/company_financials.

D'Souza, A. (2018) *Whitewalling: Art, Race and Protest in 3 Acts*, New York: Badlands Unlimited.

Davis, D.A. (2019) 'Obstetric racism: the racial politics of pregnancy, labor, and birthing', *Medical Anthropology*, 38(7): 560–573.

Deleuze, G. and Guattari, F. (1988) *A Thousand Plateaus: Capitalism and Schizophrenia*, London: Bloomsbury Publishing.

Delgado, R. and Stefancic, J. (2013) *Critical Race Theory: The Cutting Edge*, Philadelphia: Temple University Press.

Dickinson, S. (2018) 'Writing sensation: critical autoethnography in posthumanism', in S.H. Jones (ed) *Creative Selves/Creative Cultures*, London: Palgrave Macmillan, pp 79–92.

Didziokaite, G. (2017) *Mundane Self-tracking: Calorie Counting Practices with MyFitnessPal*, doctoral dissertation, Loughborough University.

Dosekun, S. (2020) *Fashioning Postfeminism: Spectacular Femininity and Transnational Culture*, Champaign: University of Illinois.

Duggan, L. (2012) *The Twilight of Equality? Neoliberalism, Cultural Politics, and the Attack on Democracy*, Boston, MA: Beacon Press.

Dyer, R. (1997) *White*, London: Routledge.

Ellis, C., Adams, T.E. and Bochner, A.P. (2011) 'Autoethnography: an overview', *Historical Social Research*, 36(4): 273–290.

Farnworth, E.R. (1999) 'Kefir: from folklore to regulatory approval', *Journal of Nutraceuticals, Functional & Medical Foods*, 1(4): 57–68.

Feagin, J.R. (2013) *The White Racial Frame: Centuries of Racial Framing and Counter-framing*, London: Routledge.

Foucault, M. (1988) 'Technologies of the self', in L. Martin, H. Gutman and P. Hutton (eds) *Technologies of the Self: A Seminar with Michel Foucault*, Amherst, MA: University of Massachusetts Press, pp 16–49.

Foucault, M. (1989) 'The ethic of the care of the self as a practice of freedom', in J. Bemauer and D. Rasmussen (eds) (1988) *The Final Foucault*, Cambridge, MA: MIT Press, pp 1–20.

Foucault, M., Davidson, A.I. and Burchell, G. (2008) *The Birth of Biopolitics: Lectures at the Collège de France, 1978–1979*, Berlin: Springer.

Fox, N.J. and Alldred, P. (2016) *Sociology and the New Materialism: Theory, Research, Action*, New York: Sage.

Fox, N.J., Bissell, P., Peacock, M. and Blackburn, J. (2018) 'The micropolitics of obesity: materialism, markets and food sovereignty', *Sociology*, 52(1): 111–127.

Friedman, M., Rice, C. and Rinaldi, J. (eds) (2019) *Thickening Fat: Fat Bodies, Intersectionality, and Social Justice*, New: York: Routledge.

Fullagar, S. (2017) 'Post-qualitative inquiry and the new materialist turn: implications for sport, health and physical culture research', *Qualitative Research in Sport, Exercise and Health,* 9(2): 247–257.

Germov, J. and Williams, L. (2017) 'Dieting women: self-surveillance and the body panopticon', in J. Sobal (ed) *Weighty Issues: Fatness and Thinness as Social Problems*, London: Routledge, pp 117–132.

Gertner, R.K. (2019) 'The impact of cultural appropriation on destination image, tourism, and hospitality', *Thunderbird International Business Review*, 61(6): 873–877.

Gill, R. and Scharff, C. (eds) (2013) *New Femininities: Postfeminism, Neoliberalism and Subjectivity*, Berlin: Springer.

Giroux, H.A. (2005) 'The terror of neoliberalism: rethinking the significance of cultural politics', *College Literature*, 32(1): 1–19.

Greenwood, B.N., Carnahan, S. and Huang, L. (2018) 'Patient–physician gender concordance and increased mortality among female heart attack patients', *Proceedings of the National Academy of Sciences,* 115(34): 8569–8574.

Grey, S. and Newman, L. (2018) 'Beyond culinary colonialism: indigenous food sovereignty, liberal multiculturalism, and the control of gastronomic capital', *Agriculture and Human Values,* 35(3): 717–730.

Griffin, M. (2012) 'Somatechnologies of body size modification: posthuman embodiment and discourses of health', doctoral dissertation, University of Central Florida, Orlando, Florida, https://stars.library.ucf.edu/etd/2387/.

Grosz, E. (2005) *Time Travels*, Durham, NC: Duke University Press.

Guthman, J. (2011) *Weighing In*, Berkeley: University of California Press.

Guthman, J. and DuPuis, M. (2006) 'Embodying neoliberalism: economy, culture, and the politics of fat', *Environment and Planning D: Society and Space*, 24(3): 427–448.

Haraway, D.J. (2004) *The Haraway Reader*, Sussex: Psychology Press.

Haraway, D.J. (2020) 'The biopolitics of postmodern bodies: constitutions of self in immune system discourse', in M. Fraser and M. Greco (eds) *The Body: A Reader*, London: Routledge, pp 242–246.

Harper, A.B. (2009) *Sistah Vegan: Black Female Vegans Speak on Food, Identity, Health, and Society*, New York: Lantern Books.

Harvey, D. (2007) *A Brief History of Neoliberalism*, Oxford: Oxford University Press.

Hayes-Conroy, J. and Hayes-Conroy, A. (2013) 'Veggies and visceralities: a political ecology of food and feeling', *Emotion, Space and Society*, 6: 81–90.

Hayles, N.K. (1993) 'Virtual bodies and flickering signifiers', *October*, 66: 69–91.

Hayles, N.K. (2008) *How We Became Posthuman: Virtual Bodies in Cybernetics, Literature, and Informatics*, Chicago: University of Chicago Press.

Heldke, L. (2001) ' "Let's eat Chinese!": reflections on cultural food colonialism', *Gastronomica*, 1(2): 76–79.

Hey, M. (2020) 'Against healthist fermentation', *Critical Dietetics*, 5(1): 12–22.

Heyes, C.J. (2006) 'Foucault goes to weight watchers', *Hypatia*, 21(2): 126–149.

Hobart, H.I.J.K. and Kneese, T. (2020) 'Radical care: survival strategies for uncertain times', *Social Text*, 38(1): 1–16.

Hooks, B. (2012) 'Eating the other: desire and resistance', in M.G. Durham and D.M. Kellner (eds) *Media and Cultural Studies: Keyworks*, New York: John Wiley & Sons, p 308.

Howell, M. (2019) 'The best kefir cultured milks and smoothies, put to the taste test', *The Telegraph*, 13 February, www.telegraph.co.uk/food-and-drink/features/best-kefir-cultured-milks-smoothies-put-taste-test/.

Johnson, E.P. (2003) *Appropriating Blackness: Performance and the Politics of Authenticity*, Durham, NC: Duke University Press.

Joshi, A. (2020) 'The cultural appropriation of turmeric', *Overachiever Magazine*, 17 April, www.overachievermagazine.com/2020/04/17/blog-post-title-one-gt6pp/.

Jovanovic, M. (2014) 'Selling fear and empowerment in food advertising: a case study of functional foods and Becel® Margarine', *Food, Culture & Society*, 17(4): 641–663.

Julier, A. (2016) 'Appropriation', in B. Cappellini, D. Marshall and E. Parsons (eds) *The Practice of the Meal: Food, Families and the Market Place*, London: Routledge, pp 101–115.

Kass, L. (1999) *The Hungry Soul: Eating and the Perfecting of our Nature*, Chicago: University of Chicago Press.

Kautz, K. and Jensen, T.B. (2013) 'Debating sociomateriality: entanglements, imbrications, disentangling, and agential cuts', *Scandinavian Journal of Information Systems*, 24(2): 89–96.

Khalikova, V.R. (2018) 'Medicine and the cultural politics of national belongings in contemporary India: medical plurality or Ayurvedic hegemony?' *Asian Medicine,* 13(1–2): 198–221.

Kimura, A.H. (2013) *Hidden Hunger: Gender and the Politics of Smarter Foods*, Ithaca: Cornell University Press.

Lau, K.L. (2016) 'Problematizing femininity in slimming advertisements', *Pertanika Journal of Social Sciences & Humanities*, 24(4): 1627–1650.

Lawson, V. (2007) 'Geographies of care and responsibility', *Annals of the Association of American Geographers*, 97(1): 1–11.

Leidy Sievert, L. (2003) 'The medicalization of female fertility: points of significance for the study of menopause', *Collegium Antropologicum*, 27(1): 67–78.

Ling, E.L.K. (2015) 'Interactive meaning potentials in weight-loss web-advertising: the female body in crisis', *Kritika Kultura,* 25: 7–25.

Loyer, J. (2017) 'The cranberry as food, health food, and superfood: challenging or maintaining hegemonic nutrition?' *Graduate Journal of Food Studies,* 4(2): 33–49.

Lupton, D. (2013) 'Quantifying the body: monitoring and measuring health in the age of mHealth technologies', *Critical Public Health*, 23(4): 393–403.

Lupton, D. (2017) *Digital Health: Critical and Cross-disciplinary Perspectives*, New York: Routledge.

Lupton, D. (2018) *Fat*, New York: Routledge.

Lupton, D. (2019a) 'The thing-power of the human-app health assemblage: thinking with vital materialism', *Social Theory & Health,* 17(2): 125–139.

Lupton, D. (2019b) 'Toward a more-than-human analysis of digital health: inspirations from feminist new materialism', *Qualitative Health Research,* 29(14): 1998–2009.

Marx, M. (1994) *Does Technology Drive History? The Dilemma of Technological Determinism*, Cambridge, MA: MIT Press.

Massey, D. (2013) *Space, Place and Gender*, New York: John Wiley & Sons.

Massumi, B. (2002) *Parables for the Virtual*, Durham, NC: Duke University Press.

McAllister, H. (2009) 'Embodying fat liberation', in E.D. Rothblum and S. Solovay (eds) *The Fat Studies Reader*, New York: New York University Press, pp 305–311.

McGillivray, D. (2005) 'Fitter, happier, more productive: governing working bodies through wellness', *Culture and Organization*, 11(2): 125–138.

McGregor, S. (2001) 'Neoliberalism and health care', *International Journal of Consumer Studies*, 25(2): 82–89.

Meskus, M. and Oikkonen, V. (2020) 'Embodied material encounters and the ambiguous promise of biomedical futures: the case of biologically derived medicines', *New Genetics and Society,* 39(4): 441–458.

Metzl, J., Kirkland, A. and Kirkland, A.R. (eds) (2010) *Against Health: How Health Became the New Morality*, New York: New York University Press.

Mikhail, G.W. (2005) 'Coronary heart disease in women is underdiagnosed, undertreated, and under-researched', *BMJ: British Medical Journal*, 331(7515): 467–468.

Nash, J.C. (2008) 'Re-thinking intersectionality', *Feminist Review*, 89(1): 1–15.

Nguyen, C.T. and Strohl, M. (2019) 'Cultural appropriation and the intimacy of groups', *Philosophical Studies*, 176(4): 981–1002.

Pausé, C. (2017) 'Borderline: the ethics of fat stigma in public health', *The Journal of Law, Medicine & Ethics*, 45(4): 510–517.

Petersen, A. and Lupton, D. (1996) *The New Public Health: Health and Self in the Age of Risk*, Thousand Oaks, CA: Sage.

Phelan, H. (2021) 'The virus, the vaccine, and the dark side of wellness', *Harper's Bazaar*, 16 March, www.harpersbazaar.com/culture/features/a35823360/covid-19-vaccine-qanon-wellness-influencers/.

Pitchbook (2021) 'Purearth', https://pitchbook.com/profiles/company/181101-70#overview.

Purearth (2021) 'Purearth', https://purearth.co.uk.

Reed-Danahay, D. (2021) *Auto/ethnography: Rewriting the Self and the Social*, London: Routledge.

Revelles-Benavente, B., Ernst, W. and Rogowska-Stangret, M. (2019) 'Feminist new materialisms: activating ethico-politics through genealogies in social sciences', *Social Sciences*, 8(11): 296, https://doi.org/10.3390/socsci8110296.

Riley, S. and Evans, A. (2018) 'Lean light fit and tight: fitblr blogs and the postfeminist transformation imperative', in K. Toffoletti, H. Thorpe and J. Francombe-Webb (eds) *New Sporting Femininities*, Basingstoke: Palgrave Macmillan, pp 207–229.

Riley, S., Evans, A. and Robson, M. (2018) *Postfeminism and Health: Critical Psychology and Media Perspectives*, London: Routledge.

Rodney, A. (2018) 'Pathogenic or health-promoting? How food is framed in healthy living media for women', *Social Science & Medicine,* 213: 37–44.

Rose, N. (2009) *The Politics of Life Itself*, Princeton, NJ: Princeton University Press.

Sanders, C. and Burnett, K. (2019) 'The neoliberal roots of modern vaccine hesitancy', *Journal of Health and Social Science*, 4(2): 149–156.

Scrinis, G. (2008) 'On the ideology of nutritionism', *Gastronomica,* 8(1): 39–48.

Scrinis, G. (2012) 'Nutritionism and functional foods', in D.M. Kaplan (ed) *The Philosophy of Food*, Berkeley: University of California Press, pp 269–291.

Selk, M.L. (2002) *Hegemonic Thinness and the Hollywood Ideal*, Berkeley: Regent University.

Shahvisi, A. (2019) 'Medicine is patriarchal, but alternative medicine is not the answer', *Journal of Bioethical Inquiry*, 16(1): 99–112.

Sikka, T. (2022) 'The social construction of 'good health', in C. Eliot and J. Greenberg (eds) *Communication and Health: Media, Marketing and Risk*, Berlin: Springer, pp 231–249.

Sikka, T. (2023) *Health Apps, Genetic Diets and Superfoods: When Biopolitics Meets Neoliberalism*, London: Bloomsbury Publishing.

Spoel, P., Harris, R. and Henwood, F. (2012) 'The moralization of healthy living: Burke's rhetoric of rebirth and older adults' accounts of healthy eating', *Health*, 16(6): 619–635.

Stibbe, A. (2004) 'Health and the social construction of masculinity in Men's Health magazine', *Men and Masculinities*, 7(1): 31–51.

Strings, S. (2019) *Fearing the Black Body*, New York: New York University Press.

TechRound (2020) 'Functional drinks brand Purearth reports 254% sales uplift since lockdown', *TechRound*, 1 October, https://techround.co.uk/news/functional-drinks-purearth-sales-uplift-lockdown/.

Turrini, M. (2015) 'A genealogy of "healthism"', *Eä-Journal of Medical Humanities & Social Studies of Science and Technology*, 7(1): 11–27.

Uotinen, J. (2011) 'Senses, bodily knowledge, and autoethnography: unbeknown knowledge from an ICU experience', *Qualitative Health Research*, 21(10): 1307–1315.

Van der Tuin, I. (2008) 'Deflationary logic: response to Sara Ahmed's imaginary prohibitions: some preliminary remarks on the founding gestures of the "New Materialism"', *European Journal of Women's Studies*, 15(4): 411–416.

Van der Tuin, I. and Dolphijn, R. (2010) 'The transversality of new materialism', *Women: A Cultural Review*, 21(2): 153–171.

Warfield, K. (2019) 'Becoming method(ologist): a feminist posthuman autoethnography of the becoming of a posthuman methodology', *Reconceptualizing Educational Research Methodology*, 10(2–3): 147–172.

White, P., Young, K. and Gillett, J. (1995) 'Bodywork as a moral imperative: some critical notes on health and fitness', *Loisir et societe/Society and Leisure*, 18(1): 159–181.

Winter, D. (2020) 'Meet the white, middle-class Pinterest moms who believe Plandemic', *The Guardian*, 6 August, www.theguardian.com/commentisfree/2020/aug/06/coronavirus-conspiracy-theories-plandemic-moms.

Wow Beauty (2018) 'Drink your way to gut health with Pure Earth kefir!' https://wowbeauty.co/health-and-wellbeing/drink-your-way-to-gut-health-with-pure-earth-kefir/.

Wrenn, C. (2019) 'Black veganism and the animality politic', *Society & Animals*, 27(1): 127–131.

Wright, J. and Harwood, V. (eds) (2012) *Biopolitics and the 'Obesity Epidemic': Governing Bodies (Vol. 3)*, New York: Routledge.

3

Ambivalent Embodiment and HIV Treatment in South Africa

Elizabeth Mills

'It's because we fought for ARVs together': HIV as a site of ongoing struggle

Miriam and I had spent the morning waiting in queues in Site C – a clinic in Khayelitsha, South Africa – as Miriam went through a series of check-ups. On this visit, the check-ups extended beyond the usual monthly blood pressure and weight checks and included more detailed tests to make sure that her HIV medicines were not having adverse effects on her body. Miriam, who had first learnt of her HIV status in 1999, was concerned that the medicines were having seriously negative effects on her body. According to Miriam, these effects – including chronic backache and weight gain – were not taken seriously by the nurses and doctors and, instead, felt she was being blamed for these conditions rather than supported in addressing them. On this visit, she had resolved to speak to the nurse and 'finally be taken seriously'. In this conversation, the nurse had told Miriam that the biomedical markers of her health – her CD4[1] count, viral load and blood pressure – indicated that the medicines were 'doing their job', and Miriam was advised to go on a stringent diet. I had waited on the bench outside the nurse's office and when Miriam came to pick me up she explained that although she was frustrated with the nurse's dismissal of her side effects, she was not surprised. With this sense of resignation, we walked across the clinic to collect her HIV medicines. While waiting in the very long queue snaking out from the three pharmacy counters to the general waiting area of Miriam's clinic, we spent most of our time chatting to the many 'comrades', as Miriam referred to them, who came up to greet her. When I commented on Miriam's active social life in the clinic, and particularly in the pharmacy queue, she smiled and said, 'Sana, it's because we fought [for ARVs] together'.

This chapter explores the ways that ARVs (antiretrovirals) are perceived and embodied by people like Miriam who had fought for the South African government to provide this treatment through the public health system. It joins other ethnographic studies that have taken place in what Moyer (2015) describes as 'the age of treatment', and follows a period of research on HIV as a global pandemic and national epidemic that, initially, could only be managed with highly expensive ARVs. Referring to a shift from ethnographic studies prior to the large-scale provision of ARVs in the 'age of AIDS', Moyer outlines a number of factors that underpinned the emergent scholarly focus on the complexities of living with ARVs:

> This rhetorical shift has been predicated on radical transformations in biomedical interventions for HIV, stemming from a combination of improved pharmaceutical treatments, massive expansion of the global health apparatus, more accurate epidemiological profiling, and a novel political framing of HIV as a disease that could be eradicated in the near future through biomedical intervention. (Moyer, 2015, p 260)

Ethnographic research in the 'age of treatment', based on a review of almost two decades (1995–2015), tends to centre on the medical management of the epidemic. Also described as 'the biomedicalisation of HIV' (Moyer, 2015, p 260), this new 'era' of HIV research has focussed on the management of HIV as a chronic disease (Torres-Cruz and Suárez-Díaz, 2020; Rosengarten et al, 2021), the social lives built around a shared identification with HIV (Valle, 2015; Girard et al, 2019) and the materiality of navigating life with HIV in discriminatory and profoundly unequal societies (Madiba and Ngwenya, 2017; Sia et al, 2020). Situated in this most recent strand of research, in the 'age of treatment', this chapter explores some of the ways that HIV-positive women experience and embody HIV and ARVs.

Although South Africa has the world's largest HIV treatment programme (UNAIDS, 2022), the narratives discussed later introduce a new set of challenges that complicate a simple reading of South Africa's HIV epidemic as manageable through biomedicine. These new challenges fan out from the historical urgency to provide ARVs, and draw in new sets of challenges around living with HIV while taking ARVs in one of the most socioeconomically unequal countries in the world (Ataguba et al, 2015; Madiba and Ngwenya, 2017; Sia et al, 2020). Working with activists like Miriam in Khayelitsha, South Africa, I found that HIV medicines not only interacted with HIV inside their bodies, but also fostered a commitment to fight together, as 'comrades', for a broader provision of HIV medicines through the public health sector. I also found that the embodiment of HIV and its medicines by the women in this research was ambivalent and shaped by the broader context in which these women navigated their lives. This chapter finds that while indisputably lifesaving, ARVs also hold a great deal of ambiguity and agency in people's lives and in their bodies. Not only do they extend life, but they also introduce risk and uncertainty and these (side) effects interact in different ways in bodies and with the social worlds that people inhabit. I discuss these linked narratives later and consider the ways in which HIV medicines have enacted a dynamic set of socialities that call attention to the world 'inside' the body as well as the precarious social and economic context in which these women live their lives.

This chapter is located at the intersection of medical anthropology and science and technology studies (STS), and it is principally informed by feminist new materialism that has brought *bios* and the materiality of our embodied lives back into focus in the humanities. In 'new materialism', '[t]he new is best understood to signal not a wholly novel moment for feminism or social theory, but rather a fresh vision of the physical and biological world, engendered through engagement with contemporary scientific fields such as quantum physics, epigenetics, and neuroscience' (Pitts-Taylor, 2016, p 4). Following Pitts-Taylor, this chapter is situated in an unfolding terrain of feminist new materialism and joins other scholars in STS to challenge representationalist accounts that rely on a strict demarcation between the social and the biological. For instance, Karen Barad's (2003; 2007; 2020) work on 'posthumanist performativity' amplifies (even politicizes) Bruno Latour's actor network theory (ANT) and his conception of non-human actants (2005), to generate findings that reveal how matter is 'an active participant in the world's becoming' (Barad, 2003, p 802). Barad's conception of post-humanist performativity builds on the foundational work of Donna Haraway and her (2006) 'cyborg theory' that, too, challenges the boundaries between human and non-human.

Across three ethnographic sections, this chapter utilizes and extends the concepts of feminist new materialism, post-humanist performativity and ANT to challenge representationalism's insistence on approaching bodies – and biologies – as uniform and passive, capable of being 'acted on' by medicines like ARVs thus rendering HIV a biomedically 'manageable' illness. I argue against treating HIV and ARVs as distinct and bounded 'actants' within abstract bodies and instead push for an understanding of all of these actants as porous and generative of each other. This argument follows Haraway's assertation that 'we are in a knot of species coshaping one another in layers of reciprocating complexities all the way down. Response and respect are possible only in those knots, with actual animals and people looking back at each other, sticky with all their muddled histories' (Haraway, 2013, p 42). The findings in this chapter complicate representationalist readings of bodies – or *bios* – as passive. Instead, they reveal the vitality of non-human actants in the lives of women like Mariam who themselves have exerted agency to access ARVs that, in turn, exert their own complicated agency within their bodies. In line with emerging research in medical anthropology and feminist new materialism (Davis, 2009; Van der Tuin, 2011; Hinton, 2014; Lupton, 2019; Fullagar and Pavlidis, 2021), this chapter's findings reflect the value of shifting away from conceiving material – like bodies, viruses and medicines – as discursively constructed, and offer a more nuanced analysis of the imbrication of human and non-human, nature and culture, science and society.

Science and the struggle for HIV treatment in South Africa

The struggle around the legitimacy of science and the efficacy of ARVs shaped South Africa's response to the emerging HIV epidemic from the late 1990s. The government's refusal to provide ARVs was initially justified on economic grounds, but this was quickly debunked and it became clear that the struggle for ARVs was an ideological one that centred on the legitimacy of science (Nattrass, 2007) or, perhaps more accurately, on the way science and medicine had been used by colonial and apartheid actors to cause harm rather than healing (Mbali, 2004; Hodes, 2018; Mbali, 2020). For over a decade, the post-apartheid government either failed to take action or refused to believe that the scientific evidence was not biased and thus perpetuated colonial and racist logics (Mbali, 2004; Chigwedere and Essex, 2010; Simelela et al, 2015). This prompted outrage and increasingly difficult questions about whose life matters in the context of a highly unequal society where very few HIV-positive people could afford to pay for a biomedical technology that would extend their life. Challenging the then president's assertion that ARVs were toxic, 'science' was invoked by activist organizations to speak to the robust evidence base which established that ARVs were vital for transforming HIV into a chronic and manageable condition. This was done in stark terms that framed science and ARVs as unambiguously efficacious (Mbali, 2004; Nattrass, 2007). However, while ARVs are indisputably effective at enabling HIV-positive people to live long lives, the black and white framing of the science by activist groups and by government officials closed down the space for recognizing some of the more nuanced and complex ways that people might experience ARVs.

South Africa's discursive struggle around the framing of science and ARVs is not unique: technological developments in HIV treatment have radically informed how activists around the world have used the discourse of science to frame access to ARVs as a human right and to call on governments to fulfil their democratic duty to provide ARVs to HIV-positive citizens. Medical anthropological theories of citizenship have flourished in the age of AIDS and of treatment, building on Adriana Petryna's (2004) original conception of biological citizenship to include, for example, health citizenship (Bänziger and Çetin, 2021), therapeutic citizenship (Nguyen, 2005; Nguyen, 2011) and pharmaceutical citizenship (Persson et al, 2016). While these are quite distinct theories based on wide-ranging ethnographic research, these conceptions of citizenship cohere around the mobilization of science to compel governments to guarantee the right to life and health for HIV-positive citizens. In South Africa, the activist struggle around the science of medicine was not simply about the mobilization of science per se (as evidence based and logical) but more specifically about the mobilization of a specific approach to the science of medicine that stripped away nuances (like side effects) and emphasized the life-enhancing capabilities of ARVs (see, for example, Nattrass, 2007). This was understandable: there was a risk that nuancing this message would result in governments, like the South African government, focussing on the risks rather than the benefits of HIV treatment and using this to justify non-provision. Nuance was dangerous when the science was relatively new, and key actors in governments were vigilant about the ways that medicine could cause harm (Ingram, 2013; Lee, 2019). In South Africa, the first decade in the 'struggle for ARVs' not only fostered but relied on an understanding of AIDS medicine as a 'technofix' (Mbali, 2004; Mbali, 2020) to support calls for the provision of ARVs through the public health sector as a human right.

While this framing of the science of medicine was vital, and central to an era-defining struggle for rights in the political history of South Africa, it also reflects the limits of representationalism. From a medical anthropological perspective, there are a range of reasons justifying a representationalist framing of medicine (conceived as 'science') as unequivocally beneficial to but also distinct from bodies (conceived of as 'natural'). For instance, Lock and Nguyen (2018) outline two assumptions that underline the premise that AIDS biomedicine is a magic bullet for restoring health and sustaining life. The first assumption is that biomedical technologies are autonomous entities. This, they argue, reinforces the construction of health-related matters as objective, quantifiable and technically manageable. Reflecting on medical anthropological research on global assemblages (Nguyen, 2005) and the imbrication of medicines and bodies within this assemblage (Cassidy, 2010; Brown et al, 2012), Lock and Nguyen challenge this assumption by arguing that 'biomedical technologies are not autonomous entities: their development and implementation are enmeshed with medical, social, and political interests that have practical and moral consequences' (Lock and Nguyen, 2018, p 1). The second assumption that Lock and Nguyen (2018) critique relates to the construction of bodies as bounded and essentially identical, which aligns with the critique offered by feminist new materialists. Not only were biomedicines left largely unexamined, and thus seen solely as a 'gold standard against which other theories of bodily affliction could be measured' (Brotherton and Nguyen, 2013, p 287), but bodies were 'black-boxed' and left by anthropologists to be probed by biologists. The relegation of bodies to 'biology' is precisely the point of concern raised by feminist new materialists, who have long sought to bring back the body – and science – into the humanities. The field of medical anthropology has also sought to disrupt false binaries like those separating science from society. Lock and Nguyen point out that this distinction was used to justify the scientific racism of the 19th and 20th centuries that equated race with biological difference and discursively located this difference in the body (Lock, 2012; Bharadwaj, 2013;, Lock and Nguyen, 2018). However, as Brotherton and Nguyen (2013) argue, the cost is that '[W]e are now blind to how social and political processes produce biological difference and by extension, how biomedical interventions may unwittingly perpetuate or enact further inequalities' (Brotherton and Nguyen, 2013, p 288).

In a powerful critique of Foucault's (1978, 1998) rather limited approach to the body, Butler's (1988, 1989) foundational writing on bodies and performativity similarly articulates the danger of separating out materiality (the body) from discourse (power). Similarly, drawing on feminist, queer, Marxist and science studies, and building on insights from Butler (1989, 2004), Foucault (1978, 1986, 1998), Hacking (1990), Rouse (2002) and others, feminist new materialist scholar Karen Barad (2007) issues a challenge to the metaphysical foundations of representationalism by offering a post-humanist performative account of the relationship between material and discursive practices (see also Barad, 2020; Barad and Gandorfer, 2021). Butler (2009), Gregson and Rose (2000) and Haraway (1997) have similarly proposed a series of conceptual approaches that incorporate performativity in order to move away from representational accounts of gender, space and science respectively. Here, I use the definition of performativity as 'the citational practices which reproduce and/or subvert discourse and which enable and discipline subjects and their performances' (Gregson and Rose, 2000, p 434).

In what follows, I draw on the notion of performativity to explore how individuals embody, reproduce and subvert power through particular sets of strategies and tactics with reference to de Certeau (1984). Integrating the conceptual approaches of performativity and intra-action (Barad, 2008, p 174), I explore how bodies become the meeting place for HIV and AIDS therapies, or non-human actants in Latour's terms (2005). I demonstrate how AIDS therapies disrupt distinctions or causal connections between the body they animate and the life they take on through this animation. As such, '[d]iscursive practices and material phenomena do not stand in a relationship of externality to one another; rather the material and the discursive are mutually implicated in the dynamics of intra-activity' (Barad, 2008, p 174). I use 'intra-action' in place of 'interaction' as the latter reflects the Newtonian legacy in which 'things', or actants, are constructed as determinant, stable, prior-existing and bounded (Barad, 2003). Intra-action, instead, reflects a feminist new materialism that seeks to understand how the materiality of things (like medicines, bodies and viruses) intra-act with discourses (like conceptions of science, medicine, health or practices of self-care, for example). This approach aligns with Barad's (2007) notion of agential realism, which emphasizes the entangled nature of objects (the realm of matter) and subjects (the realm of meaning). According to Barad, matter and meaning are entangled and it is through their entanglement – their intra-action – that they are brought into being.

In addition to working with the concepts of post-humanist performativity and intra-action, I draw on actor network theory (ANT) to think about the dynamic pathways that connect viruses with medicines in unique bodies. Specifically, I use ANT to think about the actors (people and institutions) and non-human actants (HIV and AIDS biomedicines) and how they intra-act with each other in the context of the body, and the context in which particular bodies are located. Further, I work with actor networks on the understanding that it does not put forth a theory, but rather a 'toolkit for telling interesting stories about, and interfering in, those relations' (Law, 2009, p 143). Although ANT is often discussed in abstract terms, even referred to as a theory, it is grounded in the materiality of our everyday lives. As Law writes, 'Theories usually try to explain why something happens, but actor network theory is descriptive rather than foundational in explanatory terms, which means that it is a disappointment for those seeking strong accounts. Instead it tells stories about "how" relations assemble or don't' (Law, 2009, p 141). For my research, ANT is a useful way of thinking through the kinds of people, institutions, viruses and technologies (among others) that came together and affect each other's vitality in my fieldwork. Integrating this way of thinking with my ethnographic research, I explore the intra-action of HIV and ARVs with(in) the body and describe how these actants generate different forms of risk and opportunity as they are embodied over time.

Research background

This chapter is based on multi-sited ethnography conducted between 2003 and 2016. It focusses, in particular, on ethnographic research conducted from 2011 to 2013 that traced a dynamic network of actors (activists, policy makers, healthcare systems, pharmaceutical companies) and actants (viruses and medicines) that shaped South African women's access to, and embodiment of, HIV treatment. This fieldwork included ethnographic research with a core group of women who lived in a semi-formal housing area, called Khayelitsha, in the Cape Town metropole. This group of women had played a central role in a large-scale social movement, spearheaded by the Treatment Action Campaign (TAC), that demanded the South African government provide treatment to prevent mother-to-child transmission of HIV, and highly active ARVs for the almost five million South Africans who were HIV positive (Simelela et al, 2015). The struggle for these medicines took place, broadly, from 1999 to 2009. Since then, the struggle for ARVs has changed shape but has not ended (Kaplan et al, 2017; Osler et al, 2018).

The findings drawn from my work with this core group of activists living in Khayelitsha corroborate the overall findings of the full ethnography that included: 20 key informant interviews with policy actors; 40 narrative life history interviews with men and women on antiretroviral treatment; participant photography and film; and body and journey mapping. The findings discussed in this chapter centre on field notes generated through participant observation alongside the photographs and films that were created by the group of women I worked with. The core group of women all lived in Khayelitsha, were between the ages of 30 and 50, and had all worked with TAC in various capacities in the course of the country's struggle for ARVs.

Prior to this research, I was the Deputy Director of an HIV research centre at the University of Cape Town, where I worked with policy makers, activists and academics in the field of HIV in South Africa (and Brazil). It was through this work that I had started to sense a level of concern around some of the less publicly celebrated dimensions of ARV provision, including non-adherence, side effects and treatment fatigue. I had also developed a strong set of working relationships with a number of women in this core group over the course of a decade, and I discussed working with them through a range of participatory and visual methods (including social mapping, body mapping, participatory photography and film) in order to better understand their everyday lives. Through this visual and ethnographic research, I was interested to hear about their professional work as activists while also understanding more about their embodied experiences of the medicines that they had fought to receive.

Ethics permission for this study was obtained through the University of Sussex, through consent forms with each of the study participants and through ongoing dialogue throughout the course of my research – beyond fieldwork – with the participants to discuss how they would like to be represented, which stories should be shared and which stories should not be shared at all. The accounts in this chapter reflect the first type of story.

Embodiment and new generation struggles

Historically, it is assertions of bodily sameness among AIDS activists in South Africa (Black, 2019) and Burkina Faso (Nguyen, 2010; Nguyen, 2011) that may account for successful collective action that cohered around HIV as a shared illness predicament. However, while assertions by these collectives have certainly been powerful in the history of AIDS activism, they have not accounted for the dynamic pathways that connect mutating viruses with ARVs in unique bodies that are located in shifting social, economic and political environments. In the next section, I explore how HIV and ARVs can be understood as actants with a social life. The second section focusses on the life that these actants have within the body – particularly in relation to how women experience renewed vitality alongside side effects, treatment fatigue and difficulties with adherence. The final section moves on to unpack the social and economic context in which these women live their lives, and the ways in which they work performatively with HIV and ARVs to secure scarce resources in a highly stretched economy. Through these three sections, I propose that, in understanding HIV and ARVs as non-human actants that travel complex pathways into and within women's bodies, we can also start to view some of the more nuanced dynamics that women negotiate in their everyday lives. These quotidian dynamics defy categorizations that frame HIV infection as a problem, a proxy for women's vulnerability, with ARVs cast as the 'technofix'.

The social lives of HIV and ARVs

During my fieldwork I attended a circus performance with a group of women I worked with, along with their children, as well as other young people. In the performance, Yandisa's son, Siyalela, came on stage as a doctor wielding a large syringe that represented ARVs. A performance ensued as children ran around the stage dressed up as the HIV, fooling the doctor, hurling the ARVs over trampolines and flying over each other in an attempt to stay alive and outwit the hapless ARV-wielding doctor. At the end of the circus act, the life of the virus was cut short when a brave and brazen little boy simply went up to the 'virus actors' and play-punched them in the face. This (admittedly confusing) performance of ARVs and virus actors on a stage speaks to the approach this chapter takes to explore the 'social lives' of HIV and ARVs as non-human actants with and in the arena of the body.

Our circus visit had been keenly anticipated, largely because Siyalela had been practising this scene for weeks before the circus show 'went live'. He lived with the other performers in Obs, a few houses away from uYaphi, a non-governmental organization (NGO) that described itself as an 'HIV empowerment organization'. Yandisa and six other women I came to know in my fieldwork visited uYaphi at least once a week to drop off the paper mâché bowls they had made, to collect new orders and to buy the sheets of designer paper and tubs of glue that they needed to complete the next week's work. I discuss the complex nature of this labour, under the guise of 'empowerment', later. A friend of mine who worked at the circus had been telling me about this household of young people and the playful tricks that travelled between their workplace in the circus tent in Cape Town and their home in Obs. A week before we went to the see the performance, Yandisa, over a meal of chicken and tea at *uYaphi*, expressed what she thought about this particular act, saying, 'Youth these days!' Everyone muttered in consent, as though we knew what she was talking about. I was not quite sure myself, but I thought it might have to do with young people not taking things like HIV or sex as seriously as they should. I wondered if Yandisa thought that running around in a circus (although at that stage I could not quite picture it) dressed up as the HIV was undermining the gravity of the HIV-related tuberculosis infection that, just a few years earlier, had stripped her body of all its reserves and prompted her to go back home to her mother in the Eastern Cape because she believed she was going to die. I was wrong, though. Yandisa went on to say, 'It's good, you know, that they are talking about these things. Like, well. About HIV. Because, it's a part of our life now: this virus, these drugs.'

Yandisa and all the women sitting around the table having lunch had been actively involved in the TAC's national initiative to destigmatize HIV and to compel the government to bring ARVs into the country. The women who did not work with *uYaphi* – Zama, Lilian, Sindiswa and Yvonne – had also worked as treatment literacy practitioners (TLPs) and were all at that time working in HIV-education programmes. The accounts I discuss later are grounded in TAC's treatment literacy programme, which communicated the science of HIV and ARVs to South Africans to empower them to place pressure on their doctors, local government representatives and the national government to bring ARVs into the country. What was significant in my research was the way that this scientific knowledge was internalized and complicated by the women's experiences of the effects of both HIV and ARVs as they intra-acted with each other in their bodies over time.

Barad's (2007) agential realism supports an understanding of the entanglement of HIV and ARVs. For the women that I worked with, this entanglement was visceral. As Yandisa expressed earlier, '[I]t's part of our life: this virus, these drugs.' This sentiment surfaced throughout my fieldwork and was also visible in the body maps that were created by some of the women in the core group (see for example Mills, 2017). The body maps all held various representations of ARVs and HIV, and the extent to which both actants were entangled in the women's lives. For instance, Zama drew an image of her baby in the place of her heart on her body map, and around the baby she had drawn a thick white line representing the medicines that had protected her baby from contracting HIV. Her own life, as well as her child's life, was strongly articulated in light of the power of ARVs to contain HIV and limit its agency within her and her child's bodies. In line with the critique of representationalism articulated in the introduction, these women's embodied experiences of ARVs and HIV illustrate the agency of actants, and the complex ways in which they have become 'a part of our life now'. In the two sections that follow, I explore some of the complex intra-actions of these actants in women's bodies that speak to Lock and Nguyen's (2018) caution against reading medicine as a 'techno-fix', and iterate the value of situating feminist new materialist research within very specific socioeconomic and political contexts.

'My treatment is killing me!': treatment fatigue, side effects and adherence

In moving from receiving care through the Médecins Sans Frontières (MSF) trial to regular clinical encounters, many of the women I worked with felt estranged from the care process. There were a number of reasons for this, including mistrust and fear of caregivers, and concerns around the effects of ARVs on their bodies, that they felt unable to fully communicate to clinic staff. Additionally, the discursive construction of ARVs as unequivocally beneficial by the activist organization that most of these women worked with did not offer much scope for raising concerns about negative effects of this medicine on the women's bodies. As Miriam's experience in the opening vignette demonstrates, the biomedical markers of health (like a high CD4 count, for example) precluded any further conversation about the kinds of concerns Miriam had about her chronic back ache and weight gain. The fear of ARV-related side effects and of developing resistance to HIV medicines permeated the reality of the South Africans I worked with, particularly those on second-line treatment or who had children on second-line treatment. This fear was partly founded on the fact that resistance to medicines did occur. Frequently, it occurred because of 'treatment fatigue'. In a synthesis of 21 studies that directly referred to HIV treatment fatigue, Claborn and colleagues define this phenomenon as 'decreased desire and motivation to maintain vigilance in adhering to a treatment regimen among patients prescribed long-term protocols' (Claborn et al, 2015, p 5). They note that the reasons for treatment fatigue are relatively under-researched and warrant a fuller understanding to ensure better treatment outcomes. Several of the activists I engaged with, for instance, often referred to a prominent activist who had also become tired and who had 'given up' taking medication. As the story was told, it was a shock to realize that a person who was well informed, and who worked to raise awareness around the efficacy of AIDS medicine, had reached a point where he was too 'tired' to continue with his treatment.

Treatment fatigue and viral resistance were, of course, closely linked to adherence, which in turn was shaped by people's concerns over the side effects of their treatment. Miriam, for example, frequently said 'my treatment is killing me'. One day she brought me the information sheet that was included in the tub of her Aluvia (lopinivar/ritonavir combination) tablets and showed me the section under side effects where it said that Aluvia is associated with gastrointestinal side effects. Stabbing her finger at the sheet, she asked, 'And they expect me to take this and just shut up?' Miriam also struggled with her weight (she was described by her nurse at Site C as being 'morbidly obese') and attributed this to the effects of her treatment. Miriam's frustration with her medicines was exacerbated by her sense that the healthcare workers blamed her (rather than her medicines) for being overweight and dismissed her specific health concerns, pointing instead to only the 'health' indicated by her CD4 count and her viral load. Lilian, a friend of Miriam's who lived very close to her in Khayelitsha, also experienced difficulty with a number of her ARVs and struggled to be taken seriously when raising her concerns with doctors. She explained that these fears would be sidelined by doctors who would describe them as less 'clear headed' than other more 'compliant' patients, saying, 'If you're a patient, they will take you as if you're mentally ill.'

As noted earlier, the discursive struggle around the framing of science and ARVs in South Africa played an important role in mobilizing activists to call on the South African state to provide these essential medicines through the public health system. This framing is built around a biomedical model that uses markers – like CD4 counts and viral loads – to measure health. This model positions medicines, like ARVs, as autonomous entities that work to restore health in bodies that are presumed to be bounded and essentially the same. In their critique of this model of health, medicines and bodies, Lock and Nguyen (2018) argue that biomedical interventions can unwittingly perpetuate and even enact further inequalities. To this end, the fundamental and jarring reality for people like Miriam who rely on South Africa's public health sector is that it is chronically underfunded, and that public health workers are enormously strained in their capacity to provide comprehensive care. This material reality emerges from a deeply unequal society that has a two-tier health system, with those who can afford private healthcare experiencing the impact of HIV medicines in a distinctly different way (they are able to agitate for changes to their treatment regime, for instance) compared to those who cannot afford this kind of care (and might feel like their concerns are sidelined because they are, biomedically, 'healthy') (Conmy, 2018; Winchester and King, 2018).

Many of the women I worked with had initially started taking ARVs through the MSF trial in Khayelitsha. In fact, like Zama and Miriam, these women had moved across the country to become residents of Khayelitsha in order to qualify to receive ARVs through this trial. In 2001, this was the only way that HIV-positive people could access ARVs (outside of the private health system). MSF worked closely with TAC to generate further evidence, through the trial, to compel the government to start providing this treatment through the public health system. As a result, there was a really strong sense of comradery and shared political activism between the MSF health workers and those who were receiving ARVs through the trial (Mottiar and Dubula, 2020). Miriam and Zama had first started taking ARVs through the MSF trial and they both recounted a shift in the quality of their healthcare when they moved over to receiving ARVs at their local clinic. They described how they had moved from having a relationship characterized by respect with their MSF doctor to one that was largely characterized by suspicion and mistrust by their clinic doctor. For example, although Lilian wanted to 'work with' her ARVs, and had been careful to adhere to practices of 'positive living', when her ARVs no longer worked for her she felt that she was no longer taken seriously by her doctor. She experienced health workers as punitive and explained that this has ramifications on the knowledge – and medicines – patients are able to access about and for their bodies. Expressing a concern that health workers were acting unfairly as gatekeepers to important information about medicines, and that she was required to perform a particular kind of patient 'behaviour', Lilian said:

'We make mistakes but we deserve to be treated in the right way. Like if I miss my dates, I will be shouted at by the nurses that if I miss my dates, I will die. And no one wants to be reminded that she will die. Sometimes, if the doctor finds something wrong they won't tell you, they just write it down in the folder to the pharmacist without telling you … that you must change your medication. They will just send you to the pharmacy, without telling you why you must change the medication.'

By blaming patients for noncompliance, responsibility for effective care is diverted from the caregivers to the individual patients; viral resistance and side effects are treated as symptoms of poor patient behaviour rather than as inadequate medical interventions. Lilian's sense that she would be punished for 'making mistakes' correlates with a range of studies on adherence in the 'age of treatment'. These studies similarly find that ARV programmes, while absolutely crucial, can also require a performance of neoliberal self-responsibility in order to qualify as a recipient of these medicines (Ezekiel et al, 2009; Decoteau, 2013; Kaplan et al, 2017; Cancelliere, 2020). This performance has been widely analysed in South Africa in particular and linked to the construction of the 'deserving' health citizen (Robins, 2006; Pienaar, 2016a). In her ethnographic research on TAC's promotion of health citizenship, for example, Claire Decoteau (2013) points to the dangers of placing responsibility for effective care on individuals without recognizing the structural inequalities that impact on people's experiences of medicines, and their ability to access crucial resources (like food) to be able to manage the impact on their bodies. Reflecting on Decoteau's ethnography and the insistence on a particular kind of performance of health citizenship in South Africa, Kieran Pienaar writes that:

> even those with HIV who can access these technologies and claim biomedical citizenship, 'live in a state of uncertainty, their citizenship rendered precarious and contingent, reliant on the capacity to convincingly perform "responsible" and "positive living"' (Decoteau, 2013, np). Those who fail to perform these attributes are the most impoverished South Africans, whose social marginality and material disadvantage prevents them from claiming responsibilised citizenship. (2016b)

Yvonne has changed her treatment regimen twice because of side effects, including liver damage, anaemia and lipodystrophy (the redistribution of body fat). Lipodystrophy was a commonly experienced side effect of d4T (shorthand for an antiretroviral drug called Stavudine) among the people I worked with, prior to the government's decision to remove d4T from the national treatment regimen. However, because Yvonne's biomedical 'health indicators', like her CD4 count and viral load, had remained the same, her request to change her medication was met with a threat to move her onto second-line treatment. At the time of my research, only two lines of treatment were available. Therefore, moving onto a new line of treatment would reduce the number of treatment options available to her, should she develop resistance to second-line treatment. She describes this process in relation to the struggles that are emerging with the ARV roll-out:

> 'Sometimes people get bored because it's a life-long treatment … Also side effects: I've experienced side effects on d4T. I tell myself that if the doctor doesn't want to change me, I'll stop taking ARVs … Every time I look at myself I see I'm not the same person I was before … Although my shape had changed [the doctor] told me that I need to continue with ARVs or he will put me on second line. It's as if it's only them who have information. I knew that I must change to TDF [tenofovir]. So I challenged him … and only then he changed the medication.'

Like the struggle for AIDS medicines, these emergent concerns linked to the ambivalent embodiment of ARVs continue to require strategic negotiation between healthcare workers and citizens as they navigate their right to healthcare within the public health system, which, as noted earlier, is chronically underfunded. These concerns are, however, even more complex because they call attention to the myriad effects of ARVs beyond the earlier assertions of their enactment of the 'Lazarus effect' by bringing people 'back to life' (Robins, 2006). The struggle for ARVs was a clear one: without ARVs, people would die. Now, struggles around ARVs are perhaps more difficult to mobilize around because they are embodied differently and, in a context of inequality, characterized by high levels of unemployment and economic insecurity.

The injunctions entailed in 'living positively' on ARVs have been strongly critiqued by anthropologists in ethnographies of ARV programmes for reproducing a neoliberal model of health and placing responsibility on the individual while failing to account for the complex realities of their lives (Decoteau, 2013; Copeland, 2018). As shown in Prince's (2012) ethnographic research in Kisumu, for example, these complex realities may mean that being on ARVs makes it even more difficult to live because living entails having enough money to buy food to manage taking ARVs. A negative cycle ensues, as ARVs themselves can exacerbate hunger pangs and, because they can make people quite dizzy, it becomes even more difficult for patients to work the long hours necessary to earn the money they need to buy food for themselves and their household. The biopolitical and necropolitical dynamics of accessing essential medicines like ARVs have been widely theorized and studied by medical anthropologists, but there might also be some benefit in dialoguing across disciplines and engaging with some of the questions asked 'of the body' by feminist new materialists. Josef Barla's research, for example, thinks about how bodies are policed and disciplined through biopolitics. While not explicitly looking at the provision of ARVs within a neoliberalized model of health, his work prompts questions about the way in which bodies and technologies are intimately entwined, and how bodies that 'fail' to respond 'appropriately' to technologies can be perceived as productive, as well as chaotic (Barla, 2016). The experiences recounted earlier suggest an array of intra-actions of HIV and ARVs within bodies that not only challenge biomedical models of health that position ARVs as a solution to the problem of HIV, but also productively point to the range of ways in which people embody medicines that push beyond the binary of either being healthy (and embodying technologies 'successfully') or being ill (with bodies that are 'failing').

While the provision of ARVs was perceived as absolutely critical by all the women, as discussed earlier, ARVs also appeared to be enmeshed with more complex sets of narratives. These include an ambivalence towards ARVs linked to their side effects, viral resistance and adherence. Further, as discussed later, these narratives also reflect an expansion of HIV-positive people's political relationship with the state – formerly tightly articulated around ARVs – as they called for the provision of a broader set of resources that were necessary to support a long life on ARVs. These complex sets of narratives iterate the importance of understanding the context in which bodies are located, not simply to 'situate' bodies, but as a way to more fully recognize the entanglements that move from under the skin out into the material world. Stacy Alaimo (2010, p 95), for instance, writes about the need to understand 'knowledge practices' that can show how the 'very substance of the self is interconnected with vast biological, economic and industrial systems'. This approach is also aligned with post-colonial critiques of feminist new materialism that point to the importance of understanding how structures that reinforce inequality are experienced very differently in different bodies based on their race, sex, gender, ethnicity, age, location and numerous other axes of intersectional identities (Hinton, 2014; Subramaniam et al, 2016; Cox, 2018). As Roy and Subramanian (2016, p 28) argue, 'The history of racial colonial science and medicine forcefully reminds us that we must not "decontextualize" matter from natural and cultural contexts because it is the "context" that is central to the shaping of "science" as well as to the shaping of the material "body".' The following section pays attention to these critiques by exploring the socioeconomic and political context in which women navigated their embodied experiences of HIV and ARVs.

The body in context: inequality and (un)employment

Ethnographers have drawn parallels between Rabinow's (2010) original conception of biosociality and the way that living with HIV can connect people to important resources, including medicines and support groups (Nguyen, 2005; 2010). However, my research found that when these resources dwindle, the struggle to live 'positively' on ARVs become more salient. This finding, discussed later, reflects a broader set of research conducted elsewhere in Africa, in which people on ARVs struggled to access crucial resources like food (Kalofonos, 2021) and employment (Deane et al, 2019). It was during my fieldwork between 2011 and 2013 that HIV-positive citizens started to call for the provision of a broader set of resources that were necessary to support life (over and above ARVs) (Mills, 2016b; Mills, 2017). These resources included access to employment, education, water, safe toilets and electricity and spoke to the value of understanding the socioeconomic and political contexts that are entangled with, in Alaimo's words (2010, p 95), the 'very substance of the self'.

In the early months of my fieldwork, my conversations with TAC's staff in Khayelitsha and Cape Town rippled with the news that the Global Fund had stalled on the final payment (R15 million) to the South African National Department of Health (NDoH). This, along with shifting donor priorities, had very real and negative ramifications for those working in TAC and other recipient organizations. Lilian, Yvonne, Zama and Judith, for instance, all worked in recipient HIV organizations and they had all lost their jobs by the end my fieldwork. Only Zama and Lilian subsequently found sustainable employment. These women had all been instrumental in the activist work around access to ARVs in South Africa, just as activist organizations like TAC had been instrumental in their lives and career trajectories. For example, Lilian spearheaded TAC's campaign to compel the government to provide treatment to prevent vertical transmission of HIV from mothers to babies. Both Lilian and Thandiswa testified on behalf of TAC and their affidavits were used as evidence in the court case. Lilian, like Zama, worked with an HIV treatment-literacy organization and was extensively filmed on the Siyanqoba/Beat It! television series (Hodes, 2007).

Women like Judith, Thandiswa and Zama were the first people to feel the consequences of funding restrictions and new donor priorities. Despite extensive work experience in the field of HIV activism, prevention and treatment literacy, formal education surfaced as a key factor determining their employability. Lilian and Judith also resisted the way they were typecast: not only did they resist the 'HIV-positive' label, but they resented their employers' racist and sexist behaviours. With one employer, whom they shared in the work they did on an HIV research project, they resented his assumption that they were not in a position to negotiate the terms of their employment. They were even angrier that this assumption was partly accurate because they were not in a financial position to challenge the unfair terms *or* the unfair assumptions that underpinned these terms.

These women experienced a strong lack of agency in the workplace, but they were also aware that many of these workplaces relied on them – and on their stories of violence and HIV infection – to bring in revenue. This emerged in the course of my fieldwork with six women who worked with *uYaphi*, an 'HIV empowerment' organization as noted in the earlier discussion of the circus visit. *uYaphi* set out in the late 1990s to provide employment opportunities to HIV-positive Black women on the understanding that this group was most vulnerable to the impact of HIV in South Africa. Miriam, Thandiswa, Yandisa, Zolani, Sibongile and Brenda all derived their main income through the paper mâché bowls they made for *uYaphi* each week. Each woman had joined *uYaphi* in the early 2000s through support groups that they had joined when they started ARVs.

Following a shift in management, the income generation (IG) programme became focused on selling products internationally to increase their profit margin. The IG manager explained that although the women were not given an increase in payment per product, they would benefit from increased sales by receiving larger orders. The women I worked with told me that they had not received an increase in pay for five years. According to Miriam, every time they raised this issue, the IG manager 'uses words and drawings and treats us like children but doesn't actually explain anything or listen to what we say'. Miriam became increasingly angry with *uYaphi*, and eventually challenged the manager directly and told him that he was 'exploiting my HIV'. Miriam claimed that she was punished the following week when she received an inordinately low order for bowls. Yandisa, one of the oldest women working at *uYaphi*, and a close friend (and neighbour) of Miriam's, also challenged the manager during a period of heightened contestation around wages. She asked him why a White man was running an IG programme for Black HIV-positive women; his response was that there were no suitably qualified Black women to take his position. When I saw the women after this exchange, they were both outraged by and resigned to his power. They recognized that there was no space for engagement or negotiation and were concerned about the financial implications. They did, however, irk him as often as possible by speaking about their frustrations with him in isiXhosa when he was in the same room; this frustrated him and empowered the women, because, although he heard his name, he did not speak this language and was powerless to directly challenge them as he had no basis on which to claim they were being disparaging.

Seeing the absence of state machinery in their lives, beyond their ability to access ARVs and grants, the women shifted their precarity through performance into a resource that facilitated access to the HIV activist and employment sector. I consider the way women engage with these structures as an example of performativity because they tactically navigated a set of restrictive, even oppressive, structures by subverting them. This performance was based on a strategic recognition of the development sector's construction of the HIV-positive other: the at-risk-population group of poor, Black women in South Africa. A set of struggles, resonant with Hacking's (1990) notion of a 'looping effect', proceeds from this strategic mobilization of precarity even as women resisted the label of poor, Black and HIV-positive, and came to resent the way that organizations benefited from their participation by exploiting these labels. Therefore, while all of the women I engaged with had been able to secure a degree of financial autonomy through their work in HIV activism or non-governmental organizations, they also experienced – quite directly – the impact of dwindling resources for the HIV movement. The longer-term realities of their experience of precarity in their employment speak to South Africa's historical forms of structural inequality linked to class, race and gender where women – like those I engaged with – were more likely to have been unable to complete their formal education, and therefore less likely to be able access employment opportunities, even in the informal sector. These findings also reflect the danger of operationalizing responses to the 'gender and HIV' dyad, particularly within NGOs that may 'exploit our HIV', without critical reflection about how this might reinforce rather than challenge structural inequality.

HIV, ARVs and intra-activity: towards a fuller understanding of lives in action

The timing of this ethnography, over a decade after MSF first introduced ARVs through the trial in Khayelitsha, calls into focus the trajectory of research on the complex embodiment of HIV and ARVs – both in my field site in Khayelitsha and elsewhere in Africa (Nguyen, 2011; Prince, 2012; Ataguba et al, 2015; Kaplan et al, 2017; Madiba and Ngwenya, 2017; Osler et al, 2018). This research, conducted in 'the age of treatment', speaks to the practical and discursive limits of a global and national approach that considers biomedicine as the main strategy to support people living with HIV (Florêncio, 2020; Murray et al, 2021). Also described as the 'biomedicalisation of HIV' (Pemunta and Tabenyang, 2020; Torres-Cruz and Suárez-Díaz, 2020), this approach runs the risk of reinforcing a narrow 'post-crisis' framing of HIV as a chronic but medically manageable disease. While this certainly is part of the story, the accounts of the women discussed earlier suggest that the factors necessary for living a long and full life cannot be narrowed down to the availability of ARVs. Across the three ethnographic sections, the accounts reveal a set of complex intra-actions between HIV, ARVs and bodies that are located in a context of profound socioeconomic inequality.

These ethnographic accounts reveal two main challenges experienced by women in the 'age of treatment' in South Africa. The first set of challenges is linked to the precarity engendered by AIDS biomedicines and related difficulties with side effects, adherence and viral resistance. This aspect of the new generation of struggles points to the limitations of the construction AIDS biomedicine as a 'technofix', or an autonomous entity that will have uniform effects on identical bodies. The article suggests that by understanding the dynamic intra-actions between HIV and ARVs as actants, it becomes possible to see that not only does HIV enter women's lives through human relationships, but HIV itself becomes alive and has, in Appadurai's (1988) terms, a 'social life' in the body – one that does not necessarily follow a linear trajectory that brings women's bodies from a state of ill health to one of full health. ARVs, like HIV, also call attention to the social, economic and political relationships that women navigated as activists in order to pressure the government to bring these medicines into South Africa's public health system. The range of struggles around balancing side effects with resumed biomedical health (as the nurse iterated to Miriam in the introduction) suggests that, once in the body, ARVs also take on a social life of their own. They each intra-act with each other to block HIV from proliferating. Thinking about ARVs as actants, as things with a social life, highlights another form of sociality as they become felt by the people I worked with, and visible to others, through their manifestation in side effects, like lipodystrophy. By thinking through the intra-action of HIV and ARVs as non-human actants, in line with feminist new materialism, this ethnographic research suggests that what is held in the body becomes social and, further, is shaped by the political and economic contexts in which people live.

The second set of challenges, then, relates to the social and economic contours of these women's everyday lives and the complex ways in which they worked performatively to navigate these challenges. We see, through the women's navigation of economic precarity, how HIV can be and has been used performatively to secure resources through which they were then able to navigate the tightly stretched economic landscape in which they live. Elsewhere I have described and critiqued the construction of women as vulnerable 'victims', biologically and socioeconomically more susceptible to HIV infection than men (Mills, 2016a). This chapter also considers how women work within and through this dyad to secure resources through an NGO that was specifically set up to 'empower' women. Now, however, it is the women who feel they are giving their power to the organization. Their sense that their HIV is 'exploited' runs alongside a seemingly intractable tension in which the women perform their precarity and agree to 'get with capitalism' in order to continue receiving orders for the paper mâché bowls from which they derive a crucial income for their households.

In this respect, this second challenge, around the social and economic contours of these women's lives, offers two perspectives on HIV: agency and performativity. First, it confirms the value of moving away from the 'problem/solution' framing of HIV and ARVs. This chapter shows that although HIV can be a form of embodied precarity, it can also be a resource. The second related finding, then, is that HIV – as an actant within the body – can also be performatively mobilized 'outside' the body through economic relationships to manage precarity. In this regard, the women were able to strategically secure critical economic resources in a development milieu that perpetuates the 'gender/HIV' dyad. Moreover, the women's sense of exploitation indicates the limits to structuration theory (Giddens, 1990; Giddens, 1991) as they strategically performed their precarity to secure resources while at the same time feeling that their embodied precarity – their HIV status – was a source of income, too (particularly for the NGO that was exploiting their labour and their HIV-status). Therefore, both HIV and ARVs, when thought of as 'local biologies' (Bharadwaj, 2013; Gilbert, 2013) have not only influenced the kind of social relationships that people form, but they too have social lives that are differentially embodied in people's lives.

These findings point to the value of understanding the generative dynamics between medical technologies, viruses and bodies. I suggest that this approach, which emerges through the integration of medical anthropological perspectives on embodiment and feminist new materialism, is not only theoretically valuable for bringing two disciplines into dialogue with one another but is also ontologically helpful as it more accurately reflects how people themselves experience and embody medicines and viruses in particular contexts. This approach also foregrounds an important ethical dimension about paying attention to the socioeconomic contexts that give form to certain kinds of material embodiments like vulnerability and vitality. Namely, it recognizes that inequality is unevenly distributed, moving into some bodies and lives far more deeply than others given the situated and intersectional nature of inequality. Pitts–Taylor captures this ethic succinctly here:

> While feminism's turn to matter has necessarily generated a state of reflexive retrospection about its theories of sex/gender, the biological body, and the mediations of science, the aim is not really more nuanced, abstract theoretical perspectives, the overcoming of epistemic impediments to grasping nature, nor even the ontology of matter per se. Rather it is the concrete, particular, and situated lives of beings as they are caught up in the workings of power/knowledge/ontology. (Pitts–Taylor, 2016, p 16)

Writing about the ethics of new materialism, Haraway (2008) and Alaimo (2008) similarly reflect on the ethical value of bringing *bios* and the 'fleshiness of bodies' back into focus in the wake of the constructivist 'turn'. For instance, Alaimo writes that 'Acknowledging that one's body has its own forces, which are interlinked and continually intra-acting with wider material as well as social, economic, psychological and cultural forces, can not only be useful, but may also be ethical' (Alaimo, 2008, p 250). The ethnographic research in this chapter highlights the extent to which bodies have their own forces that are in an ever unfolding and dynamic union with other forces, like HIV, medicines, economic structures and political histories. This acknowledgement of intra-acting forces within particular socioeconomic and political contexts is not only ethical, but ontologically vital: it is, to put it simply, a way of seeing the fullness of lives in action and in context.

Conclusion

Despite the ethical and ontological implications of research on the intra-activity of HIV and ARVs, there is still a wide gulf between the field of medical anthropology and the practice and funding of global health programmes. For instance, global organizations that guide funding priorities and national policy approaches continue to advocate for measures that are predicated on the provision of treatment, without sufficient attention to the range of concerns that people experience when taking ARVs. Not only is this framing problematic, largely because of its narrow focus on the biomedicalization of HIV, but it also reinforces a 'post crisis' framing of a pandemic that is very far from over. For instance, the United Nations (UN) has called for ambitious benchmarks in HIV treatment, testing and adherence through its '90–90–90' testing and treatment cascade. This cascade was initiated by UNAIDS in 2016 with a view to bringing the 'end of AIDS' by ensuring that 90% of people living with HIV knew their status and, of this group, 90% were able to access treatment and that, of this group, 90% had undetectable viral loads. The UN noted in 2020, however that it was unlikely that the target would be met (UNAIDS, 2020a). This 'endgame' framing is ambitious, but it is also cravenly accompanied by a flat-lining of global HIV funding for people living with HIV and the organizations that support them. This framing precipitates a paradox: wealthy countries in the global North and powerful organizations like the UN generate a post-crisis discourse ostensibly to advocate for a radical 'end to AIDS'. This discourse is framed as ambitious, but it bypasses the very real struggles that continue to be faced by far less wealthy countries, and far less powerful organizations, that know that the struggle to 'end AIDS' is far from over.

This 'post crisis' framing also results in substantial global disparities in the management and control of the pandemic – one that transects scale, influencing national budgets, local health providers and individual lives (Dalton, 2018; Powers, 2019; Murray et al, 2021). Ethnographic research in Mozambique, for instance, explores how household impoverishment intersects with a national health system that has been stripped of resources through national and international neoliberal austerity measures. Looking specifically at maternal and infant ARV provision, Chapman (2021) found that HIV-positive women and healthcare workers faced similar struggles in managing profound scarcity in their homes and clinics. Attempts to manage this scarcity – through delayed initiation on ARVs, intermittent adherence and low retention in ARV programmes – compromised the health of women and infants, and undermined the overall outcomes of the ARV programme. The very real implications of living with HIV, and taking ARVs, while navigating entrenched socioeconomic inequalities have been documented in ethnographic research across Southern Africa (and indeed throughout the world). Taken together, this research, along with the findings of my own fieldwork discussed previously, highlights the harm implicit in the post-crisis framing of HIV (Kenworthy et al, 2018). Therefore, while research on the biomedicalization of HIV has generated a rich understanding of the complexities of living with HIV, the 'post crisis' framing (at a time of global austerity) occludes the significant challenges that continue to be faced by HIV-positive people, and by HIV-positive women in particular.

Note

[1] A CD4 cell count test is a laboratory test that measures the number of CD4 T-cells in a cubic millimetre of blood. A CD4 count is an indicator of immune function in patients living with HIV.

Acknowledgement

This chapter draws on fieldwork that is included in the author's monograph, published with Bristol University Press, entitled *HIV, Gender and the Politics of Medicine: Embodied Democracy in the Global South*.

References

Alaimo, S. (2008) 'Trans-corporeal feminisms and the ethical space of nature', in S. Alaima and S. Hekman (eds) *Material Feminisms*, Bloomington, IN: Indiana University Press, pp 237–264.

Alaimo, S. (2010) *Bodily Natures: Science, Environment, and the Material Self*, Bloomington, IN: Indiana University Press.

Appadurai, A. (1988) *The Social Life of Things: Commodities in Cultural Perspective*, Cambridge: Cambridge University Press.

Ataguba, J. E.-O., Day, C. and Mcintyre, D. (2015) 'Explaining the role of the social determinants of health on health inequality in South Africa', *Global Health Action*, 28865.

Bänziger, P.-P. and Çetin, Z. (2021) 'Biological citizenship and geopolitical power play: health rights of refugees living with HIV in Turkey', *Critical Public Health*, 31: 43–54.

Barad, K. (2003) 'Posthumanist performativity: toward an understanding of how matter comes to matter', *Signs*, 28: 801–831.

Barad, K. (2007) *Meeting the Universe Halfway: Quantum Physics and the Entanglement of Matter and Meaning*, Durham, NC: Duke University Press.

Barad, K. (2008) 'Living in a posthumanist material world – lessons from Schrodinger's cat', in A. Smelik and N. Lykke (eds) *Bits of Life: Feminism at the Intersections of Media, Bioscience, and Technology*, Seattle: University of Washington Press.

Barad, K. (2020) 'After the end of the world: entangled nuclear colonialisms, matters of force, and the material force of justice', *Estetyka i Krytyka*, 58: 85–113.

Barad, K. and Gandorfer, D. (2021) 'Political desirings: yearnings for mattering (,) differently', *Theory & Event*, 24: 14–66.

Barla, J. (2016) *Technologies of Failure, Bodies of Resistance: Mattering*, New York: New York University Press.

Bharadwaj, A. (2013) 'Subaltern biology? Local biologies, Indian odysseys, and the pursuit of human embryonic stem cell therapies', *Medical Anthropology*, 32: 359–373.

Black, S.P. (2019) *Speech and Song at the Margins of Global Health: Zulu Tradition, HIV Stigma, and AIDS Activism in South Africa*, New Brunswick, NJ: Rutgers University Press.

Brotherton, P.S. and Nguyen, V.-K. (2013) 'Revisiting local biology in the era of global health', *Medical Anthropology*, 32: 287–290.

Brown, T., Craddock, S. and Ingram, A. (2012) 'Critical interventions in global health: governmentality, risk, and assemblage', *Annals of the Association of American Geographers,* 102: 1182–1189.

Butler, J. (1988) 'Performative acts and gender constitution: an essay in phenomenology and feminist theory', *Theatre Journal,* 40: 519–531.

Butler, J. (1989) 'Foucault and the paradox of bodily inscriptions', *The Journal of Philosophy,* 86: 601–607.

Butler, J. (2004) *Undoing Gender,* New York; London, Routledge.

Butler, J. (2009) *Performativity, Precarity and Sexual Politics*, Antropólogos Iberoamericanos en Red (AIBR).

Cancelliere, F. (2020) 'The politics of adherence to antiretroviral therapy: between ancestral conflicts and drug resistance', *DADA Rivista di Antropologia Post-globale,* 2: 13–41.

Cassidy, R. (2010) 'Global expectations and local practices: HIV support groups in the Gambia', *AIDS care,* 22: 1598–1605.

Chapman, R.R. (2021) 'Therapeutic borderlands: austerity, maternal HIV treatment, and the elusive end of AIDS in Mozambique', *Medical Anthropology Quarterly,* 35: 226–245.

Chigwedere, P. and Essex, M. (2010) 'AIDS denialism and public health practice', *AIDS and Behavior,* 14: 237–247.

Claborn, K.R., Meier, E., Miller, M.B. and Leffingwell, T.R. (2015) 'A systematic review of treatment fatigue among HIV-infected patients prescribed antiretroviral therapy', *Psychology, Health & Medicine,* 20: 255–265.

Conmy, A. (2018) 'South African health care system analysis', *Public Health Review,* 1: 1–8.

Copeland, T. (2018) 'To keep this disease from killing you: cultural competence, consonance, and health among HIV-positive women in Kenya', *Medical Anthropology Quarterly,* 32: 272–292.

Cox, L. (2018) 'Decolonial queer feminism in Donna Haraway's "A Cyborg Manifesto" (1985)', *Paragraph,* 41: 317–332.

Dalton, D. (2018) 'Cutting the ribbon? Austerity measures and the problems faced by the HIV third sector', in P. Rushton and C. Donovan (eds) *Austerity Policies*, Berlin: Springer, pp 173–195.

Davis, N. (2009) 'New materialism and feminism's anti-biologism: a response to Sara Ahmed', *European journal of Women's Studies,* 16: 67–80.

Deane, K., Stevano, S. and Johnston, D. (2019) 'Employers' responses to the HIV epidemic in sub-Saharan Africa: revisiting the evidence', *Development Policy Review,* 37: 245–259.

De Certeau, M. (1984) *The Practice of Everyday Life*, Berkeley: University of California Press.

Decoteau, C. (2013) *Ancestors and Antiretrovirals: The Biopolitics of HIV/AIDS in South Africa,* Chicago: University of Chicago Press.

Ezekiel, M.J., Talle, A., Juma, J.M. and Klepp, K.-I. (2009) '"When in the body, it makes you look fat and HIV negative": the constitution of antiretroviral therapy in local discourse among youth in Kahe, Tanzania', *Social Science & Medicine,* 68: 957–964.

Florêncio, J. (2020) 'Antiretroviral time: gay sex, pornography and temporality "post-crisis"', *Somatechnics,* 10: 195–214.

Foucault, M. (1978) *The History of Sexuality (Vol. 1),* New York: Pantheon.

Foucault, M. (1986) 'The use of pleasure', in *The History of Sexuality (Vol. 2),* New York: Pantheon.

Foucault, M. (1998) 'Technologies of the self', in L. Martin (ed) *Technologies of the Self,* Amherst, MA: University of Massachusetts Press, pp 16–49.

Fullagar, S. and Pavlidis, A. (2021) 'Thinking through the disruptive effects and affects of the coronavirus with feminist new materialism', *Leisure Sciences,* 43: 152–159.

Giddens, A. (1990) *The Consequences of Modernity,* Cambridge: Cambridge University Press.

Giddens, A. (1991) 'Structuration theory: past, present and future', in C. Bryant and C. Jary (eds) *Giddens' Theory of Structuration: A Critical Appreciation,* London: Routledge, pp 210–221.

Gilbert, H. (2013) 'Re-visioning local biologies: HIV-2 and the pattern of differential valuation in biomedical research', *Medical Anthropology,* 32: 343–358.

Girard, G., Patten, S., Leblanc, M.A., Adam, B.D. and Jackson, E. (2019) 'Is HIV prevention creating new biosocialities among gay men? Treatment as prevention and pre-exposure prophylaxis in Canada', *Sociology of Health & Illness,* 41: 484–501.

Gregson, N. and Rose, G. (2000) 'Taking Butler elsewhere: performativities, spatialities and subjectivities', *Environment and Planning D,* 18: 433–452.

Hacking, I. (1990) *The Taming of Chance – Ideas in Context,* Cambridge: Cambridge University Press.

Haraway, D. (1997) *Modest_Witness@Second_Millennium: FemaleMan_Meets_Oncomouse: Feminism and Technoscience,* New York, Routledge.

Haraway, D. (2006) 'A cyborg manifesto: science, technology, and socialist-feminism in the late 20th century', in J. Weiss, J. Nolan, J. Hunsinger and P. Trifonas (eds) *The International Handbook of Virtual Learning Environments,* Berlin: Springer, pp 117–158.

Haraway, D. (2008) 'Otherworldly conversations, terran topics, local terms', *Material Feminisms,* 3: 157.

Haraway, D.J. (2013) *When Species Meet*, Minneapolis: University of Minnesota Press.

Hinton, P. (2014) ' "Situated knowledges" and new materialism(s): rethinking a politics of location', *Women: A Cultural Review,* 25: 99–113.

Hodes, R. (2007) 'HIV/AIDS in South African documentary film, c. 1990–2000', *Journal of Southern African Studies,* 33: 153–171.

Hodes, R. (2018) 'HIV/AIDS in South Africa', *Oxford Research Encyclopedia of African History,* www.academia.edu/37285882/HIV_AIDS_in_South_Africa.

Ingram, A. (2013) 'After the exception: HIV/AIDS beyond salvation and scarcity', *Antipode,* 45: 436–454.

Kalofonos, I. (2021) *All I Eat Is Medicine: Going Hungry in Mozambique's AIDS Economy,* Berkeley, CA: University of California Press.

Kaplan, S.R., Oosthuizen, C., Stinson, K., Little, F., Euvrard, J., Schomaker, M. et al (2017) 'Contemporary disengagement from antiretroviral therapy in Khayelitsha, South Africa: a cohort study', *PLoS Medicine,* 14: e1002407.

Kenworthy, N., Thomann, M. and Parker, R. (2018) 'From a global crisis to the "end of AIDS": new epidemics of signification', *Global Public Health,* 13: 960–971.

Latour, B. (2005) *Reassembling the Social: An Introduction to Actor-Network-Theory,* Oxford: Oxford University Press.

Law, J. (2009) 'Actor network theory and material semiotics', in B.S. Turner (ed) *The New Blackwell Companion to Social Theory,* Chichester: Wiley-Blackwell, pp 141–158.

Lee, R. (2019) 'Art, activism and the academy: productive tensions and the next generation of HIV/AIDS research in South Africa', *Journal of Southern African Studies,* 45: 113–119.

Lock, M. (2012) 'The epigenome and nature/nurture reunification: a challenge for anthropology', *Medical Anthropology,* 32: 291–308.

Lock, M.M. and Nguyen, V.-K. (2018) *An Anthropology of Biomedicine,* New York: John Wiley & Sons.

Lupton, D. (2019) 'Toward a more-than-human analysis of digital health: inspirations from feminist new materialism', *Qualitative Health Research,* 29: 1998–2009.

Madiba, S. and Ngwenya, N. (2017) 'Cultural practices, gender inequality and inconsistent condom use increase vulnerability to HIV infection: narratives from married and cohabiting women in rural communities in Mpumalanga province, South Africa', *Global Health Action,* 10: 1341597.

Mbali, M. (2004) 'AIDS discourses and the South African state: government denialism and post-apartheid AIDS policy-making', *Transformation: Critical Perspectives on Southern Africa,* 54: 104–122.

Mbali, M. (2020) 'From AIDS to cancer: health activism, biotechnology and intellectual property in South Africa', *Social Dynamics,* 46: 449–470.

Mills, E. (2016a) '"When the skies fight": HIV, violence and pathways of precarity in South Africa', *Reproductive Health Matters,* 24: 85–95.

Mills, E. (2016b) '"You have to raise a fist!" Seeing and speaking to the state in South Africa', *IDS Bulletin,* 47(1).

Mills, E. (2017) 'Biopolitical precarity in the permeable body: the social lives of people, viruses and their medicines', *Critical Public Health,* 27: 350–361.

Mottiar, S. and Dubula, V. (2020) 'Shifting consciousness and challenging power: women activists and the provision of HIV/AIDS services', *Law, Democracy and Development,* 24: 158–176.

Moyer, E. (2015) 'The anthropology of life after AIDS: epistemological continuities in the age of antiretroviral treatment', *Annual Review of Anthropology,* 44: 259–275.

Murray, D.A.B., Benton, A., Graham, J., Hassan, W., Jolly, J., Lorway, R. et al (2021) *Living with HIV in Post-Crisis Times: Beyond the Endgame,* Lanham, MD: Lexington Books.

Nattrass, N. (2007) *Mortal Combat: AIDS Denialism and the Struggle for Antiretrovirals in South Africa,* Pietermaritzburg: University of KwaZulu-Natal Press.

Nguyen, V.-K. (2005) 'Antiretroviral globalism, biopolitics and therapeutic citizenship', in A.O.A.S.J. Collier (ed) *Global Assemblages: Technology, Politics and Ethics as Anthropologica Problems,* Malden, MA: Blackwell Publishing.

Nguyen, V.-K. (2010) *The Republic of Therapy*, Durham, NC: Duke University Press.

Nguyen, V.-K. (2011) 'Trial communities: HIV and therapeutic citizenship in West Africa', in P.W. Geissler and C. Molyneux (eds) *Evidence, Ethos and Experiment: The Anthropology and History of Medical Research in Africa,* Oxford: Berghan Books, pp 429–444.

Osler, M., Hilderbrand, K., Goemaere, E., Ford, N., Smith, M., Meintjes, G. et al (2018) 'The continuing burden of advanced HIV disease over 10 years of increasing antiretroviral therapy coverage in South Africa', *Clinical Infectious Diseases,* 66: S118–S125.

Pemunta, N.V. and Tabenyang, T.C.J. (2020) 'The sociocultural context of HIV/AIDS in the Eastern Cape region', in N.V. Pemunta and T.C.J. Tabenyang (eds) *Biomedical Hegemony and Democracy in South Africa.* Leiden: Brill.

Persson, A., Newman, C.E., Mao, L. and De Wit, J. (2016) 'On the margins of pharmaceutical citizenship: not taking HIV medication in the "treatment revolution" era', *Medical Anthropology Quarterly,* 30: 359–377.

Petryna, A. (2004) 'Biological citizenship: the science and politics of Chernobyl-exposed populations', *Osiris,* 19: 250–265.

Pienaar, K. (2016a) 'Conclusion: towards an ontological politics of disease', in *Politics in the Making of HIV/AIDS in South Africa*, Basingstoke: Palgrave Macmillan, pp 120–134.

Pienaar, K. (2016b) *Politics in the Making of HIV/AIDS in South Africa*, Basingstoke: Palgrave Macmillan.

Pitts-Taylor, V. (2016) 'Feminism, science, and corporeal politics', in V. Pitts-Taylor (ed) *Mattering,* New York: New York University Press, pp 1–20.

Powers, T. (2019) 'Echoes of austerity: policy, temporality, and public health in South Africa', *Focaal,* 2019: 13–24.

Prince, R. (2012) 'HIV and the moral economy of survival in an East African city', *Medical Anthropology Quarterly,* 26: 534–556.

Rabinow, P. (2010) 'Artificiality and enlightenment: from sociobiology to biosociality', *Politix,* 90(2): 21–46.

Robins, S. (2006) 'From "rights" to "ritual": AIDS activism in South Africa', *American Anthropologist,* 108: 312–323.

Rosengarten, M., Sekuler, T., Binder, B., Dziuban, A. and Bänziger, P.-P. (2021) *Beyond Biological Citizenship: HIV/AIDS, Health, and Activism in Europe Reconsidered,* New York: Taylor & Francis.

Rouse, J. (2002) *How Scientific Practices Matter: Reclaiming Philosophical Naturalism,* Chicago: University of Chicago Press.

Roy, D. and Subramaniam, B. (2016) 'Matter in the Shadows', in V. Pitts-Taylor (ed) *Mattering,* New York: New York University Press, pp 23–42.

Sia, D., Tchouaket, É.N., Hajizadeh, M., Karemere, H., Onadja, Y. and Nandi, A. (2020) 'The effect of gender inequality on HIV incidence in sub-Saharan Africa', *Public Health,* 182: 56–63.

Simelela, N., Venter, W.F., Pillay, Y. and Barron, P. (2015) 'A political and social history of HIV in South Africa', *Current HIV/AIDS Reports,* 12: 256–261.

Subramaniam, B., Foster, L., Harding, S., Roy, D. and Tallbear, K. (2016) 'Feminism, postcolonialism, technoscience', in E. Hackett, O. Amsterdamska, M. Lynch, J. Wajcman and W. Bijker (eds) *The Handbook of Science and Technology Studies,* Cambridge, MA: MIT Press, p 407.

Torres-Cruz, C. and Suárez-Díaz, E. (2020) 'The stratified biomedicalization of HIV prevention in Mexico City', *Global Public Health,* 15: 598–610.

UNAIDS (2020a) '90–90–90: good progress, but the world is off-track for hitting the 2020 targets', www.unaids.org/en/resources/presscentre/fea turestories/2020/september/20200921_90–90–90.

UNAIDS (2020b) 'Country Factsheets: South Africa', www.unaids.org/en/regionscountries/countries/southafrica.

Valle, C.G.D. (2015) 'Biosocial activism, identities and citizenship: making up "people living with HIV and AIDS" in Brazil', *Vibrant: Virtual Brazilian Anthropology,* 12: 27–70.

Van Der Tuin, I. (2011) 'New feminist materialisms', *Women's Studies International Forum,* 34: 271–277.

Winchester, M.S. and King, B. (2018) 'Decentralization, healthcare access, and inequality in Mpumalanga, South Africa', *Health & Place,* 51: 200–207.

4

An 'Artificial' Concept as the Opposite of Human Dignity

Kazuhiko Shibuya

Introduction

This chapter provides a critical review of recent developments of science and technology, and scrutinizes the challenges of *artificiality* that we invent. Human existence and dignity continue to be challenged by advanced technologies such as AI and other innovations in the life sciences. As such, I argue that frameworks emerging out of science and technology studies (STS), ethics, philosophy and ELSI (ethical, legal and social implications) must be used to inform our perspectives.

I begin by focusing on the concept of the 'artificial'. When one adds the prefix 'artificial' to any word, it introduces a level of discomfort, and it feels as if it conflicts with the values associated with life and human existence. As discussed further on, it carries with it some elements of the 'uncanny valley' or 'numinöse'. We should thus also feel a sense of latent warning rather than admiration for technological triumph, particularly when confronted with the nature of 'artificiality'.

Second, based on the artificial as a concept, I argue that human existence and dignity are at risk, due to the development of specific forms of science and technology. Despite this, human beings and our societies still continue to pursue the mass production and purchase of these artefacts. What kind of society will we produce in the future? Will the end point of technological innovation mean the loss of human dignity and the value of life itself? Will humanity have the will to control the advanced technologies and artefacts we created?

This chapter aims to engage in a critical review of technological trends. It offers an overview of the values that really need to be protected and draws on a metaphysical perspective from which to maximize sustainability and existential survival into the future.

Problems with the 'artificial' concept

Humanity can be described, at least in part, as a population of mammals who produce their own technology and tools (for example, *Homo faber* (Scheler, 1961)). 'Artificiality', which is different from natural origin, can be generally defined as a something human-caused and human-made, and refers to those *artefacts* produced since the beginning of recorded civilization (Needham,1956; Jones and Taub, 2018). Therefore, human history does not intersect perfectly with natural history. Human history can be said to be a process of human beings becoming independent from nature, unveiling natural phenomena and using these principles to artificially replace natural things with artificial ones (what we now call technologies).

Historically, according to the pioneering discussions in Aristotle's *Physics* and other texts, technologies are artefacts that can be regarded as an imitation of nature, constituted by *hyle* (material, matter) and *eidos* (form). However, this ancient framework does not readily apply to modern research such as virtual projection and other digital information technologies. The concept of artificiality, insofar as human beings imitate nature, has been understood on this basis in many disciplines. There have also been interdisciplinary debates about 'artificiality' (Simon, 1996) in STS, such as that about the human versus the non-human (Latour, 1988; 2008), and the social constructivism of facts and technological artefacts (Pinch and Bijker, 1984).

Two meanings of 'artificial'

Today, any phrase including the prefix 'artificial' (for example, *artificial* reality, *artificial* intelligence, *artificial* arm (limbs) (McGee and Maguire, 2007), *artificial* organs, *artificial* hearts and so on) contains ambiguity in terms of its meanings.

First, it conveys sense of technological achievement and invention by humanity, and it denotes a declaration of victory that humanity has brought natural phenomena 'under control'. It also claims artificial equivalency and superiority for a natural product or system. For example, searching on the term 'artificial' on the National Institutes of Health PubMed Central,[1] which is one of the world's largest medical databases, returns technical terms related to medicine such as artificial cell (in a simple query, more than 612,347 relevant papers were listed, as of November 2021), artificial organ (more than 211,054), artificial skin (more than 167,716) and human artificial chromosome (more than 105,902), as well as terms related to interdisciplinary themes between medicine and engineering, such as artificial intelligence (AI) (more than 91,720).

Second, in opposition to such a positive descriptor, 'artificial' can also connote a level of immorality, one that refers to a deviation from nature as well as the 'forbidden' and 'taboo'. The depiction of human-like androids, for example, invokes some discomfort – such a feeling is called 'uncanny valley'. Of course, the 'artefacts' produced by humanity itself are not inherently accompanied by negative meanings. However, because of the unique way in which they diverge from natural phenomena, various risks caused by advanced technologies have produced a variety of concerns (Feenberg, 2015).

The current increase in the number of concepts and technologies with the prefix 'artificial' demonstrates that the nature of human existence has been called into question (Latour, 2013). This remains the case even though the number of such artefacts on the Earth has been increasing since the birth of our civilization. According to Elhacham et al (2020), the 'anthropogenic mass' has exceeded the mass of naturally occurring materials. Human civilization has already filled the Earth with innumerable artefacts for our convenience. Living in the age of the Anthropocene[2] means that the earth is being gradually replaced by such anthropogenic mass and artificial objects to such an extent that it will eventually become an *artificial* planet.

Feeling latent risks and threats towards the 'artificial'

Certainly, there are many products that seem essential and difficult to replace. However, there are also many that make us feel an uneasiness towards the prefix 'artificial'. Try adding the prefix to any life-related event or concept. For example, artificial love, artificial placentas, artificial infants, artificial mothers, artificial families, artificial lovers, artificial siblings, artificial animals, artificial plants, artificial foods (artificial apples, artificial grapes, artificial bread, artificial wine and so on) and others ... Instantly, there is sense of crossing the line that must not be crossed. Further, there is also the controversial issue of genetic engineering of newborns by genome editing (Cyranoski, 2019), and if restrictions are not properly implemented, both the *mother's womb* and the *newborn* may be degraded to 'artificial' objects.

This clearly shows that a feeling of taboo towards a life-related concept with an 'artificial' prefix is linked to the erosion of the dignity of life and crossing an irreversible threshold. Because our technology imitates nature, modern research and development (R&D) in business is unconsciously transforming existential beings with the highest value into something that has lost its meaning. This loss is rooted in the violation of the dignity of life – a perceptual violation and one rooted in a religious perspective.

First, bioethical issues regarding the modification of life (Mulkay, 1994) have been debated for decades (for example, the Asilomar principle). Even today, there tends to be a deep-rooted aversion to the artificial modification of life (for example, genetic modification or genome editing). According to a comprehensive survey in Europe[3] (Bauer and Gaskell, 2002) and a similar case in Japan (Shibuya, 2013), aversion to genetic modification technology implies an attitude that is based on emotional disgust rather than citizens' scientific understanding. And yet, it was decided that an edited 'tomato'[4] would be available on the Japanese market by the end of 2020. Ethical debate, including whether or not to call such food an 'artificial tomato' will continue to occur (for example, collective symbolic coping with genetic modified tomato (Wagner and Kronberger, 2001), tomato modified by gene-editing technology (Zsögön et al, 2018)).

Second, from a psychological point of view, a type of anxiety disorders known as *phobias*, especially, *technophobia* become relevant (Weil and Rosen, 1995; David, Lee-Kelley and Barton, 2013). Based on perceptual information, artificial and technological objects make people feel a specific kind of discomfort. The greater the discrepancy between people's sense of reality and their represented perception (Yuste et al, 2017), the more difficult it becomes for people to agree and accept such objects. On the other hand, the closer the object is to human existence, the more it can threaten identity[5] and dignity. People will sense the potential threats posed by such objects (that is, uncanny valley). The image of artificially compounded 'chemical substances', for example, has also been characterized as too 'negative'[6] by social surveys in Japan. One of the reasons is believed to be the serious effects of pollution on the environment and concerns around human health. In other words, citizens' attitudes towards these chemical risks are not only based on their level of understanding of natural science, but also on psychological factors. In fact, natural life (including humankind, and other animals) can be regarded as chemical compounds, but the public's attitudes suggest chemical substances = artefacts = undesirable. This is not self-denial but, rather, a mental aversion to artificial manipulation and incomprehensible things.

Third, there are issues that conflict with religious values. Religious values and cultural customs are often incompatible with technological development, and tough reactions can be expected when technological innovation overturns traditional values about human existence. Another negative emotion is based on 'Numinöse'. As Otto argued in his book *The Idea of the Holy* (1923), it is an awe and unfathomable fear that comes from the 'creature consciousness' of the Creator (that is, the God), and is connected to an attitude of self-discipline or religious identity towards taboos that must not be violated. In particular, the 'shuddering mystery' represents an attitude of awe and fear of something divine. In this context, when we feel discomfort towards 'artefacts', it might invoke the fear of violating divinity. This recalls one of the philosophical issues in Kant's *Critique of Pure Reason*, which is a question rooted in the boundary between morality and religion: 'What shall we not do?' According to Jung, 'Numinöse' is one of the archetypes hidden in the unconscious that unveils diverse symbols of culture and society, interweaving with our mentality in depth (Jung, 1969). From their point of view, such an emotional sensation towards artefacts may be a psychological response revealing a moral barrier common to all human beings.

Of course, the problem is not only that artefacts make us feel uncomfortable. It is natural to feel that latent *risks* to the dignity of life are distorted by such manipulations. This sense is based on several additional factors, including whether or not the manipulation is temporary, whether or not it is a lifelong process, and whether the effects will be passed on to subsequent generations. This includes concerns about biological engineering through genetic modification and genome editing, which would threaten the dignity of not only specific individuals but, ultimately, all of humanity.

Points to be discussed around AI and bioengineering

Next, I discuss four issues around artificiality, human existence, science and technology including the roles of *dignity*, *controllability*, *legal authority* and *value*. Finally, I discuss the essential issues of human society arising from advanced science and technology such as AI and the life sciences (Owe and Baum, 2021).

Dignity

First, let's re-examine the issue of dignity. Lexically, dignity is rooted in the German concept of death. The verb used to describe the death of a dignified human being is *sterben*, while the verb used to describe the death of an animal-like existence is *verenden* (Heidegger, 2008). In brief, depending on which verb is used for the 'death' of an android as one of the artefacts, it should be questioned whether it is a dignified being or not. Similarly, the semantic difference between 'being' and 'existence' can be summed in English as 'being with consciousness or not'. As such, the relationship between dignity and autonomous consciousness turns out to be extremely important.

In this regard, there are already ample philosophical criticisms of consciousness and AI (Searle, 1980; Dreyfus and Dreyfus, 1986). Recent engineering research and development in AI using human neurons[7](Hassabis et al, 2017) and *artificial* consciousness[8] is well advanced (Buttazzo, 2001; Haladjian and Montemayor, 2016). We need to consider whether robots should be considered 'beings' with dignity (Chrisley, 2008).

Philosophical theories of human beings from a mechanistic point of view, such as in de la Mettrie's book *L'homme-machine*, Descartes's dualism of mind and body, Spinoza's philosophical monism, Wiener's cybernetic theory of human beings and the recent debates on the philosophy of the brain have all dealt with such issues on a theoretical basis. However, in the current situation, where the level of technology that these theories have deliberated is advancing more swiftly than expected in the medical field, engineering research such as BMI (brain–machine interface), bionic[9] systems and cyborgs has also been progressing (Zeng, Sun and Lu, 2021). Replacing missing parts of the human body for disabled people, surgical procedures to enhance the lives of healthy people and the replacement of various organs with artificial ones and reactive sensors have all become possible. Further, various types of robots and sensing systems have been developed based on bio-inspired mechanisms and morphology (Contreras-Vidal and Grossman, 2013). These have led to the mass production of robots that look like pseudo-life forms. These discussions are clearly related to a series of technologies referred to as 'human augmentation' and 'body augmentation' (Agar, 2013). While some have had positive consequences, including assistance and training for the elderly and the disabled (Donati et al, 2016; Dawson et al, 2021), and improving symptoms in severely depressed patients by artificially stimulating their brains (Scangos et al, 2021), these technologies will not always have foreseeable outcomes.

What these technologies have in common is that they artificially modify human existence and its modality. The problem here centres on the fact that the boundaries between humans and artificial robots and surrounding human dignity (Von Braun et al, 2021) have become blurred. The fact that artificial modification can be intentionally achieved implies a mechanical manipulability of both the body and the mind.

As one indicator, the presence or absence of own consciousness is still important vis-à-vis its ability to discern many objects and life forms in nature as well as mechanical artefacts. Yet, if humanity artificially creates one of these 'beings with artificial consciousness' (for example, androids), dignity does not necessarily accompany it. If it does, it means that dignity is mass-producible. Thus, it seems that there are elements other than 'consciousness' that are lacking in order to be considered equivalent to humans. To be sure, there are various discussions and diverse definitions of human dignity (Schroeder et al, 2017), and a modality of this dignity can be determined by 'inviolable value'. Importantly, the value of natural beings strictly depends on its own uniqueness and the irreversibility of life phenomena – both of which are *sui generis* (Shibuya, 2020; 2022). Principally, I have pointed out that human dignity can be derived from the highest good (*summum bonum*), that it relies on the historical uniqueness of living existence and is rooted in the physical and mental integrity of each individual. It also speaks to the significance of genetic inheritance, the irreversibility and limitation of one's own lifespan and the unity of philosophical existence. In the future, in any transformation to a state where our existence can be identified as mass-producible, artefacts such as artificial robots will inevitably undermine what it means to be human.

For example, there is a metaphor in ancient Greek mythology called 'Theseus' ship'. It poses the question whether, if there is an old ship in which broken pillars and planks are replaced with other materials, the final ship can be said to be the same as the original ship. Similarly, when we replace the structure of the human body with 'artefacts', as mentioned earlier, at what point can we say that we are the same person? Up to what point can we say that human dignity and identity are preserved (Shibuya, 2020; 2022)? Could a 'human' who had been completely replaced with mechanical parts like a robot be said to be a human with dignity?

Controllability

A second issue of concern is whether humans themselves have complete control over the artificial objects they produce. In particular, there are persistent opinions that AI-driven androids will become a threat to humanity. Therefore, managing artificial products, namely, the reliability of the AI, is all-important. Further, there is also the problem of *the ubiquity of AI* in Cyber Physical Systems[10] (Rahwan et al, 2019). If these systems become the foundation of human society, the question is one of how humanity could oversee and manage such systems as a whole.

An illustrative example, is the 'Ring of Gyges' problem, which draws out questions of ethics associated with Plato's introduction of the ancient Greeks. This is an ancient Greek story in which a man named Gyges obtained a magic ring which made him invisible when he put it on. He began to play all sorts of pranks with it and ended up killing a king and usurping his throne. In light of this story, ethics poses the question of whether or not people act morally when others are not looking. Applying this to the trustworthiness of AI, AI is constituted by software and computational processes, the consequences of which we cannot always foresee. Yet it can be important to anticipate some pretty dangerous situations – for example, the weaponization of AI. To do so, various arguments[11] exist for accountable AI as Explainable AI (XAI) – but in essence, accountability depends on the human ability to verify (Cath, 2018; METI, 2021; Wing, 2021).

As such, humanity must not relinquish its ultimate authority of control over artificial products. It is important to point out that the essence of such concerns is not so much about AI itself as it is about the humans who produce and use it. We need a reflexive view around self-regulation so as to not to violate our dignity through the use of technology.

In addition, advanced science and technology also requires legal, ethical and religious frames in its governance (Hagendorff, 2020). For example, the pope is concerned about AI, and in fact the content of his November 2020 'prayer intention' was about 'artificial intelligence'.[12] However, the question of whether or not humans as described earlier have dignity, and whether or not androids created by bioengineering technologies such as genome editing can be considered human, will continue to bring up ethical and religious questions. Will 'artificial newborns' from 'artificial wombs' have the dignity of human beings? Are artificial androids made up of bioengineered 'artificial brains' and 'artificial biological tissues' human beings?

Even beyond these limitations, when advanced technology eventually reaches the point of artificially reproducing human consciousness, humans will produce artificial beings that are equivalent to us in intelligence (for example, android robots). Will such artefacts equipped with artificial consciousness follow Plato's 'three parts of the soul' (λόγος = Logos, θυμός = Thymos, ἐπιθυμητής = Desire (for example, eros))? The more perfectly equipped they are with all three parts, the more human-like their existential modality will be. Given that it is difficult for humans to maintain cooperative relationships with each other (ÓhÉigeartaigh et al, 2020), there is no reason to assume that there will always be peace between humans and robots with artificial consciousness, or that robots with artificial consciousness will always remain subordinate to humans in the absence any regulations and moral mechanisms.

Thus, it is time for us to reconsider where the technological innovations with elemental technologies described as 'artificial' are headed.

Legal authority

Third, concerns must be raised for privacy, patents and other rights in relation to bioengineering and AI (Jaynes, 2021). This is because these technologies risk violating the grounds of unique human identity and reflect a host of other ethical and legal risks (Knoppers and Greely, 2019; Shibuya, 2020). The first is the problem of the existential genetic basis of human beings, and the second is the current situation where the boundary between the existential and virtual modality of humankind is becoming blurred due to the progress of information technology.

First, for example, a specific gene and chemical substance can be patented if it is found to be useful and novel for pharmaceutical and medical purposes. This raises quite serious concerns about the ability to patent humans and life forms per se.

Genes that control specific functions in an organism, or medicinal chemicals that promote functions, are also subject to patenting, whether for medical or economic purposes. These should be subject to improper use restrictions and patenting rights, even though the highest good and dignity are often excluded from economic valuations. As General Data Protection Regulation (GDPR) legislation (Slokenberga, Tzortzatou and Reichel, 2021) notes, and including genome data, it is the free and inherent right of every citizen to retain and properly control of his or her own body, which is fundamental to democracy, human rights and dignity (for example, Uyghur issue,[13] COVID-19 related issues; Shibuya, 2006; 2021c[14]).

Second, these issues also highlight the problem of legal action and human existence at the boundary between real and virtual persons in the case of *deep-fake* technology and AI deep learning. Human rights regarding ethnic, racial, gender and other minorities in relation to deep-faking are also being challenged, as well as fairness in relation to AI machine learning (Kamishima et al, 2012). The artificial falsification of videos not only creates a false image but also distorts the personality rights of the person on whom the video is based. From the point of view of moral rights as a part of copyright, it is not impossible, but not sufficient, to legally protect against criminal acts arising from deep-fake technology (Shibuya, 2021a; Nema, 2021). It has also become possible for AI to completely create an original virtual person (avatar[15]) that does not actually exist, and to use such virtual persons for advertising and broadcast commentary videos. Who will be legally responsible for any damage caused by the avatars created by AI?

The issue of legal rights related to the existence of human beings is becoming more complex in legal interpretation with the development of such technologies. According to Japan's copyright law, moral rights are clearly defined as a part of copyright, which is the general term for the right to protect the creator of a work from being mentally injured. Works of art, literature, music and video made by me reflect my thoughts and feelings. However, depending on international treaties and the copyright laws of each country, the extent to which artificial products are specifically recognized as an infringement of moral rights differs. In particular, in the digitized society, those rights will be litigated, and legal action should be mandatory in the future over the unauthorized processing of electronic personal information.

The unconscious transformation of value

Fourth, the pros and cons of the conversion of human dignity to economic value need to be examined from other perspectives as well.

This includes valuation issues related to Rawls's discussion of the fairness principle (1971, 2001). Rawls presents the arguments that a greater burden should be placed on the able-bodied in order to equalize the load between them and the disabled. However, since many physical disabilities are based on genetic heredity, is it fair if the difference in inherent value is genetically determined (Harden, 2021)? If the genetic qualities of each individual engender statistically significant differences in socioeconomic opportunities between citizens, is it fair to intentionally equalize the talent, appearance and socioeconomics of each individual by manipulating genes? And, is it also ethical to *artificially* manipulate this distribution of characteristics (so called *artificial* justice and fairness)? What these issues mean is that while there are serious cases where there is absolute observance of human dignity in its ethical and moral dimensions, some cases are treated with relative valuation. Which is to say that there can be decision making that demands absolute adherence to the highest good of the individual, while at the same time judging the value of an individual's existence based on utilitarian priorities and valuation criteria in society as a whole.

In other words, the perspective that the dignity of human existence is fixed and eternally unshakable may not be valid. Rather, we must realize that the evaluation of even the highest good is relative. The factors that bring about such moral relativity depend mainly on the observer as well as social and public criteria. When the highest good of individual origin is arranged in such a way, the priority and the criteria for evaluating value as the subjective view of the observer naturally intervenes in the individual's highest good. Thus, although the comparison of the highest good of each individual might not always be the most important, the relative superiority or inferiority of dignity is determined by someones' intention. It could be a doctor doing triage, a couple trying to apply genome editing to have their ideal child, a human resources manager, an adversary or even an AI.

Further, the question of how AI will analyse and manage personal health data, including genetic information, will become even more problematic in the future (Mello and Wang, 2020; Cenci and Cawthorne, 2020). When a meritocratic perspective is allowed to intervene in the highest good of human beings, evaluation by AI will also be calculated by a similar algorithm to determine aspects of human existence (Shibuya, 2020; 2022). The worry is that the value of human existence as an absolute will be relativized and eventually reduced to the facts deprived of even a concept of its dignified value.

Crossing the worlds

In fact, the artificiality and its problems that I have discussed has been already engaged in national projects. The Japanese government's Sixth Basic Plan for Science, Technology and Innovation[16] includes an item (one of the goals of the moonshot R&D system) to 'realize a society in which people are free from the constraints of body, brain, space and time by 2050'. As described earlier, when I append an extra axis of *real* and *virtual* to the *natural–artificial* dichotomy, we can see that artificial cyberspace is likely to undermine the natural realm. Natural (natural origin):

> Real:
>> Existential existence (life forms)
>> Genetic functions
>> Neural circuits in brain

Virtual:
> Mental image and representation in brain
> Possible Worlds (for example, modal logics)
> Worldviews based on quantum theory

Artificial (anthropogenic origin):

Real:
> Technologies
> Artefacts (for example, robots, tools, and so on)
> Life manipulation (for example, gene modification, gene editing, induced pluripotent stem (iPS) cells, and so on)

Virtual:
> Algorithm-driven entity (for example, software-based AI, agent and so on)
> Big Data
> VR (virtual reality), AR (augmented reality), XR (extended reality)
> Metaverse.

Examining these elements of real and virtual realms, these differences are deeply related to the identity of humanity, such as whether or not it involves one's own physical or existential corporeality, whether or not it is a naturally occurring life form (that is not anthropogenic), the degree of invasion of the body, whether or not changes are reversible, and the congruence between the physical body and consciousness (Shibuya, 2020). People identify own meaning and ascribe value to objects whether these are artefacts or not. In such virtual spaces, all people are managed as digital data of 0 or 1. And since numbers have no meaning, it is presumed that everyone will be treated equally and fairly. However, I don't presume that certain attributes related to health, gender, race, ability and origin will be accepted as a positive marker of individuality.

Good and evil are the result of man-made actions. For example, Microsoft's Tay over-learned malicious data from external users, which led it to make discriminatory remarks against minorities and races. It is the responsibility of humanity to properly educate the AI as to the differences between good and evil. AI and androids are artefacts that are unlikely to be legally regarded as natural persons based on natural law (*lex naturalis*). Thus, androids do not retain human rights (so called *artificial* sovereignty) (Gordon and Pasvenskiene, 2021). In the first place, the natural law is strictly founded on God, Nature, or rationality based on human nature (for example, Kelsen, Hobbes and others). Thus, there is room for debate as to whether or not we accept that the AI should be assigned an *artificial* authority to govern the state and citizens (Guice, 1998) and a legal right to control the causal relationships of complex things and events in society.

Further, we should consider what to do if such an artificial world becomes pervasive not only for each individual but also for society as a whole. As discussed in this chapter, there are many arguments that support this position (Anderson and Anderson, 2021; Falco et al, 2021), including the *singularity* of the AI and bioengineering and the various risks it may pose to human existence (Bostrom, 2013; Bostrom and Ćirković, 2018; Vladimir and Jean-Philippe, 2021). As objects currently fill the Earth, there is no assurance that human beings will be able to keep the natural world under their control through manipulation and scientific methods.

Given that AI-related technologies and high-performance computing (and quantum computing in the future) will be more advanced than they are now, the worry is that humanity may become subordinate to the *artificial* world realized by these technologies. It is also necessary to reconsider whether the AI-driven systems controlling everything from people's genome information to behavioural patterns will be a utopia or dystopia. Also of concern is whether, when our modality and value of existence are changed by those technologies, the meaning of our life itself will also be redefined in those contexts.

Towards a sustainable future for humanity

Role of academic research

Aristotle contemplated the essence of the artificial concept from philosophy, in line with the level of science and technology at that time. Similarly, in order to delve into the effects of modern science and technology today, it is necessary for many disciplines derived from philosophy to critically analyse artificiality from their own perspectives. With the development of information science such as AI and bioengineering, science and technology policy needs to be interrogated more comprehensively, including from humanities and social sciences – particularly *design*, *engineering*, and *management*. These fields are best placed to deal with, for example, controversies around on designer babies and global problems in public health (for example, countermeasures in a corona pandemic). Sensitive issues related to human existence and sustainable survival need to be examined comprehensively and critically.

The first is design, which is includes systematization and quantitative analysis of social economy, and the mathematical and scientific research necessary to keep them in an optimal state (Douglas, Howard and Lacey, 2021). Design has various contexts (Vardouli, 2015), and includes the social sciences in general (economics, law, politics, and so on), medical economics and policy, mathematical modelling, and fields related to engineering design

(Akrich, 1992). Product design and science fiction prototyping from an artistic point of view can also be considered (David, 2020).

The second category is the disciplines directly related to engineering production and manufacturing processes. Many engineering disciplines fall into this category, and information systems, products, and buildings are produced in this way.

Third, are fields that include risk theory, science and technology policy theory, research evaluation, and policy science in general. For example, the discipline of regulatory science has already been proposed in Europe. It is known to deal with issues such as chemical substances, environmental problems, medical care and public health (Guerra, 2021). In cases where risk is a concern, but decision making and policy formulation based on scientific evidence is not possible, business and policy implementation should be based on safety (Baker-Brunnbauer, 2021).

Last, but not least, philosophy and STS (Latour, 1987, 2005) would critically commit to above three parts. A mechanism to verify, feed back, and control the degree to which human dignity and bioethical perspectives are met in design and engineering can be carried out in research in the field of management (Coeckerlbergh, 2020).

Prioritizing risk reduction and sustainable development goals

Why is it that humanity tends to prioritize science and technology that is difficult to control, with consequences that threaten its own survival, instead of ones that are relevant to survival? While the promotion of state power, the military and industry is given priority, technologies related to minorities and human rights tend to be put on the back burner. Science and technology policy is state policy, and its nature is no different whether the state is democratic or despotic. We must also address R&D progress from the perspective of STS and the sociology of science (Woolgar, 1991), in particular the structure of professional groups (for example, scientists, government officials, engineers) involved in science and technology, as well as the public understanding of science and public participation in science.

Since ancient Greece, the study of the elements has been a research directly related to the understanding of natural objects and their artificial applications. For example, artificial nucleosynthesis (such as synthetic element 113: *nihonium*[17]) by scientists was carried only in pursuit of achieving basic research, and not focused on commercialization (Shibuya, 2007). The success of artificial elements, artificial photosynthesis[18] and others was the result of the unceasing efforts of scientists in Japan. In this way, the curiosity-driven pursuit of natural laws and principles is only in order to achieve academic goals, while the development of engineering technology is to establish a pipeline for business success based on the basic laws. Such business goals seek and prioritize profit first, and do not usually focus on the survival of human society or compliance with ethical values. As proof of this, the process of innovation that produces some technology rarely involves first creating the means to invalidate such a technology before developing it further.

This often leads to a situation where a certain engineering technology causes a problem, and then we are forced to conduct R&D to fix it (Winner, 1986). In the case of nuclear technology (nuclear fusion[19] (the 'artificial sun'), and nuclear fission) (Jasper, 2014), the control mechanisms for extracting energy have been researched and developed, but why haven't the technologies for safely disposing of nuclear materials? Shouldn't the order of development be reversed in the first place? Shouldn't we be promoting nuclear energy research and development only *after* we have developed technologies that can neutralize the effects of nuclear materials?

In Japan, ten years after the Fukushima[20] nuclear power plant disaster of 2011, not even the debris has been recovered, and the radiation has not been safely blocked (NRC, 2011; OECD/NEA, 2016).

Citizens' and consumers' risk assessment of nuclear damage and scepticism about food safety in Fukushima were not the same as that by scientists, engineers and business (Science Council of Japan, 2011; Shibuya, 2020; 2022). In the author's field survey and interviews (Shibuya, 2017; 2018; 2021b), there were a number of cases where consumers who suspected radiation contamination avoided purchasing agricultural products (their safety was scientifically confirmed) that were labelled as originating in Fukushima Prefecture. Many narratives from interviewees clearly indicated the asymmetry of information between sellers and buyers regarding safety and the different standards for risk assessment among the stakeholders. The following is an excerpt from an interview.

'Just because it was produced in Fukushima Prefecture, it was sold at a low price in the market. Rice and other products were bought cheaply for business use, and profits are low. Anyway, the situation is still now terrible. From the buyer's point of view, there is no need to be particular about rice produced in Fukushima (rice can be produced anywhere), so it can be cheaper ... We have received compensation, especially from TEPCO [an electric power company, the company that caused the nuclear disaster in Fukushima], but only for shipping restrictions. It's not about human damage or psychological compensation. We're trying to do group claims in court.'

In Fukushima, the entire local community, including the elderly, women and children, suffered devastating damage. This catastrophic event exemplifies that humanity is not equipped technologically to solve a nuclear disaster. These technologies tend to accelerate in the direction that brings more profit rather than the protection of human dignity. Technology assessment is of course quite important. But, as responsibility in uncertain situations and the precautionary principle show, it is not only the prioritization of technology development, but also the development of science and technology against unexpected events that is of concern.

The same concerns about various risks apply to all of the latest technologies as well as AI and bioengineering (Coeckelbergh, 2010; Roeser et al, 2012). Artefacts of AI and bioengineering operate under the same logic as nuclear technology development, since nobody can nullify these technologies yet. The impact of biohazards and environmental destruction caused by technologies will continue to affect humanity. Similarly, AI-equipped robots and products also need to be properly controlled, repaired and disposed of if problems occur after mass production.

A certain logic on the part of businesses and industries often hinders these concerns. Even if there is a risk that a certain technology will adversely affect humanity, it is generally profitability that determines whether a company will invest. Such business-centred logic on the part of the companies is also in line with the logic of national governments trying to foster domestic industry. In universities and research institutes, ideas that can be patented or turned into businesses are given top priority. As a result, innovations that nullify or dispose of the negative effects of technology may not be prioritized for commercialization. Corporate logic is reluctant to develop such disabling and safety technologies unless there is a firm prospect of matching supply and demand. Without some kind of legal governance and ethical management in place, AI, robotics and bioengineering technologies developed by such companies could lead to the mass production of 'runaway cars with no brakes'. It will be too risky for the nation and humanity if the technologies that correspond to the brakes are considered only after the target technologies have caused problems.

As discussed throughout this chapter, one of the answers to the question of why the potential threats of and fears about artefacts have not disappeared seems to be largely due to the fact that the development of such technologies prioritizes profit, neglects risk assessment from the perspective of the consumers and does not establish disabling technologies aimed at the target technology. The development of safe disposal methods and the establishment of technologies that impose less of a burden on the environment must also be recognized as an important part of research. It's not about one company monopolizing the profits; rather, we should turn towards technologies that benefit humanity and society as a whole. Indeed, the design and engineering divisions of academic research can be considered as the gas pedal in an automobile. The management division can be expected to function as the steering wheel and the brake, and to control humanity's choice of course and speed of progress as appropriate. If such institutional mechanisms are not functioning and there is no self-reflection, at the present breakneck speed of development if advanced technologies, humanity will be in a very precarious position.

Conclusion

In this chapter I have critically discussed various significant issues in science and technology based on the concept of artificiality. The history of science and technology reflects the growth of humanity's understanding of natural phenomena and their modification. The concept of artificiality is symbolic of this history. Technological innovation will continue to engender a qualitatively new reality – one that we have never before experienced. Further research related to science and technology needs to continue to scrutinize trends in technology as they are produced by humanity. Here, I have pointed out the pros and cons of value transformation around human dignity, which will require focused discussion in the future, particularly around AI and the life sciences. No matter what the consequences of the technologies for our social worlds, we must be willing to take on more responsibility. This is because humanity has no choice but to *artificially* create a safe, sustainable and ideal world for ourselves and for future generations.

Notes

[1] www.ncbi.nlm.nih.gov/pmc/

[2] www.nature.com/news/polopoly_fs/1.17085!/menu/main/topColumns/topLeftColumn/pdf/519144a.pdf

[3] https://gmoinfo.eu/eu/articles.php?article=Most-Europeans-hardly-care-about-GMOs&mc_cid=add66ba0b0&mc_eid=822a149de0

[4] https://asia.nikkei.com/Business/Science/Japan-approves-gene-edited-super-tomato.-But-will-anyone-eat-it

[5] www.nature.com/articles/d41586-019-03014-4

[6] www.nite.go.jp/data/000010206.pdf (in Japanese)

[7] www.csail.mit.edu/news/new-deep-learning-models-require-fewer-neurons

[8] www.jst.go.jp/kisoken/crest/en/project/1111083/15656376.html

[9] www.science.org/journal/scirobotics

[10] These are systems that collect diverse data in the real world (physical space) through sensor networks and integrate them in cyberspace. In particular, from the perspective of the ubiquity of AI, with the development of Internet of Things and 5G communication networks, there is concern that AI will govern the organization of the society.

[11] https://digital-strategy.ec.europa.eu/en/policies/expert-group-ai

[12] 'Pope Francis urges followers to pray that AI and robots "always serve mankind"', *The Verge*, www.theverge.com/2020/11/11/21560076/pope-francis-ai-for-good-pray-serve-mankind

[13] www.nature.com/articles/d41586-019-03775-y

[14] www.nature.com/collections/diedggejjf/

[15] www.soulmachines.com/, https://youtu.be/rjclQ3m5JRw?t=122

[16] https://www8.cao.go.jp/cstp/stmain/mspaper3.pdf

[17] www.riken.jp/medialibrary/riken/pr/publications/anniv/riken100/part1/riken100-1-5-5.pdf (in Japanese).
[18] www.riken.jp/medialibrary/riken/pr/publications/riken88/riken88-2-5.pdf (in Japanese).
[19] www.iter.org/
[20] www.nature.com/collections/hlslgwwzkl

References

Agar, N. (2013) *Truly Human Enhancement: A Philosophical Defense of Limits*, Cambridge, MA: MIT Press.

Anderson, S.L. and Anderson, M. (2021) 'AI and ethics', *AI and Ethics*, 1: 27–31, https://doi.org/10.1007/s43681-020-00003-6.

Akrich, M. (1992) 'The de-scription of technical objects', in W. Bijker and J. Law (eds) *Shaping Technology, Building Society: Studies in Sociotechnical Change*, Cambridge, MA: MIT Press.

Baker-Brunnbauer, J. (2021) 'Management perspective of ethics in artificial intelligence', *AI and Ethics*, 1: 173–181.

Bauer, M.W. and Gaskell, G. (2002) *Biotechnology: The Making of a Global Controversy*, Cambridge: Cambridge University Press.

Bostrom, N. (2013) 'Existential risk prevention as global priority', *Global Policy*, 4(1): 15–31.

Bostrom, N. and Ćirković, M.M. (eds) (2018) *Global Catastrophic Risks*, Oxford: Oxford University Press.

Buttazzo, G. (2001) *Artificial Consciousness: Utopia or Real Possibility? IEEE Computer*, https://ieeexplore.ieee.org/abstract/document/933500/.

Cath, C. (2018) 'Governing artificial intelligence: ethical, legal and technical opportunities and challenges', *Philosophical Transactions Royal Society A (Mathematical, Physical and Engineering Sciences)*, 376: 20180080, https://doi.org/10.1098/rsta.2018.0080.

Cenci, A. and Cawthorne, D. (2020) *Refining Value Sensitive Design: A (Capability-Based) Procedural Ethics Approach to Technological Design for Well-Being, Science and Engineering Ethics*, https://doi.org/10.1007/s11948-020-00223-3.

Chrisley, R. (2008) 'Philosophical foundations of artificial consciousness', *Artificial Intelligence in Medicine*, 44(2): 119–137.

Coeckelbergh, M. (2010) 'Engineering good: how engineering metaphors help us to understand the moral life and change society', *Science and Engineering Ethics*, 16: 371–385.

Coeckelbergh, M. (2020) *Introduction to Philosophy of Technology*, Oxford: Oxford University Press.

Contreras-Vidal, J.L. and Grossman, R.G. (2013) 'NeuroRex: a clinical neural interface roadmap for EEG-based brain machine interfaces to a lower body robotic exoskeleton', 35th Annual International Conference of the IEEE Engineering in Medicine and Biology Society (EMBC), pp 1579–1582, https://doi.org/10.1109/EMBC.2013.6609816.

Cyranoski, D. (2019) 'China to tighten rules on gene editing in humans', *Nature*, https://doi.org/10.1038/d41586-019-00773-y.

David, A. (2020) 'Science and culture: "design fiction" skirts reality to provoke discussion and debate"', *Proceedings of the National Academy of Sciences,* 117(24): 13179–13181, https://doi.org/10.1073/pnas.2008206117.

David, G., Lee-Kelley, L. and Barton, M. (2013) 'Technophobia, gender influences and consumer decision-making for technology-related products', *European Journal of Innovation Management,* 6(4): 253–263.

Dawson, J., Liu, C., Francisco, G., Cramer, S., Wolf, S., Dixit, A. et al (2021) 'Vagus nerve stimulation paired with rehabilitation for upper limb motor function after ischaemic stroke (VNS-REHAB): a randomised, blinded, pivotal, device trial', *Lancet*, 397(10284): 1545–1553, https://doi.org/10.1016/S0140-6736(21)00475-X.

Donati, A., Shokur, S, Morya, E., Campos, D., Moioli, R., Gitti, C. et al (2016) 'Long-term training with a brain-machine interface-based gait protocol induces partial neurological recovery in paraplegic patients', *Scientific Reports,* 6: 30383, https://doi.org/10.1038/srep30383.

Douglas, D.M., Howard, D. and Lacey, J. (2021) 'Moral responsibility for computationally designed products', *AI and Ethics,* 1: 273–281.

Dreyfus, H.L. and Dreyfus, S. (1986) *Mind over Machine: The Power of Human Intuition and Expertise in the Era of the Computer*, New York: Free Press.

Elhacham, E. et al (2020) 'Global human–made mass exceeds all living biomass', *Nature,* 588: 442–444, www.nature.com/articles/s41586-020-3010-5.

Falco, G., Shnneiderman, B., Badger, J., Carrier, R., Dahbura, A., Danks, D. et al (2021) 'Governing AI safety through independent audits', *Nature Machine Intelligence,* 3: 566–571.

Feenberg, A.L. (2015) 'Critical theory of technology and STS', https://doi.org/10.3390/isis-summit-vienna-2015-T1.0.1003.

Gordon, J.S. and Pasvenskiene, A. (2021) 'Human rights for robots? A literature review', *AI and Ethics,* 1: 579–591.

Guerra, G. (2021) 'Evolving artificial intelligence and robotics in medicine, evolving European law: Comparative remarks based on the surgery litigation', *Maastricht Journal of European and Comparative Law,* https://doi.org/10.1177/1023263X211042470.

Guice, J. (1998) 'Controversy and the state: Lord ARPA and intelligent computing', *Social Studies of Science,* 28(1): 103–138.

Hagendorff, T. (2020) 'The ethics of AI ethics: an evaluation of guidelines', *Minds and Machines*, 30: 99–120.

Haladjian, H.H. and Montemayor, C. (2016) 'Artificial consciousness and the consciousness–attention dissociation', *Consciousness and Cognition,* 45: 210–225.

Harden, K.P. (2021) *The Genetic Lottery: Why DNA Matters for Social Equality*, Princeton, NJ: Princeton University Press.

Hassabis, D., Kumaran, D., Summerfield, C. and Botvinick, M. (2017) 'Neuroscience-inspired artificial intelligence', *Neuron*, 97: 245–258.

Heidegger, M. (2008) *Sein und Zeit (English translated edition)*, New York: Harper Perennial.

Jasper, J.M. (2014) *Nuclear Politics: Energy and the State in the United States, Sweden, and France*, Princeton, NJ: Princeton University Press.

Jaynes, T.L. (2021) 'On human genome manipulation and homo technicus: the legal treatment of non-natural human subjects', *AI and Ethics*, 1: 331–345.

Jones, A. and Taub, L. (2018) *The Cambridge History of Science (Vol. 1–8)*, Cambridge: Cambridge University Press.

Jung, C.G. (1969) *The Archetypes and the Collective Unconscious (English translated edition)*, Princeton, NJ: Bollingen Foundation.

Kamishima, T., Akaho, S., Asoh, H. and Sakuma, J. (2012) 'Fairness-aware classifier with prejudice remover regularizer', in M.-R. Amini, S. Canu, A. Fischer, T. Guns, P.K. Novak and G. Tsoumakas (eds) *Machine Learning and Knowledge Discovery in Databases*, New York: Springer, pp 35–50.

Knoppers, B.M. and Greely, H.T. (2019) 'Biotechnologies nibbling at the legal "human"', *Science*, 366(6472): 1455–1457.

Latour, B. (1987) *Science in Action: How to Follow Scientists and Engineers through Society*, Cambridge, MA: Harvard University Press.

Latour, B. (1988) 'Mixing humans and nonhumans together: the sociology of a door-closer', *Social Problems*, 35(3): 298–310.

Latour, B. (2005) *Reassembling the Social: An Introduction to Actor-Network-Theory*, Oxford: Oxford University Press.

Latour, B. (2008) 'When things strike back: a possible contribution of "science studies" to the social sciences', *The British Journal of Sociology*, 51(1): 107–123.

Latour, B. (2013) *An Inquiry into Modes of Existence: An Anthropology of the Moderns*, Cambridge, MA: Harvard University Press.

McGee, E.M. and Maguire, G.Q. (2007) 'Becoming Borg to become immortal: regulating brain implant technologies', *Cambridge Quarterly of Healthcare Ethics*, 16(3): 291–302.

Mello, M.M. and Wang, C.J. (2020) 'Ethics and governance for digital disease surveillance', *Science*, 368(6494): 951–954.

METI, Japan (2021) 'AI Governance in Japan Version 1.0 (Interim Report)', www.meti.go.jp/press/2020/01/20210115003/20210115003-3.pdf.

Mulkay, M. (1994) 'The triumph of the pre-embryo: interpretations of the human embryo in parliamentary debate over embryo research', *Social Studies of Science*, 24(4): 611–639.

Needham, J. (1956) *Science and Civilisation in China (Vol. 1, Introductory Orientations)*, Cambridge: Cambridge University Press.

Nema, P. (2021) 'Understanding copyright issues entailing deepfakes in India', *International Journal of Law and Information Technology,* eaab007, https://doi.org/10.1093/ijlit/eaab007.

NRC (Nuclear Regulatory Commission, USA) (2011) *The Near-term Task Force Review of Insights from the Fukushima Dai-Ichi Accident,* http://pbadupws.nrc.gov/docs/ML1118/ML111861807.pdf.

OECD / NEA (2016) *Five Years after the Fukushima Daiichi Accident: Nuclear Safety Improvements and Lessons Learnt,* www.oecdnea.org/nsd/pubs/2016/7284-five-years-fukushima.pdf.

ÓhÉigeartaigh, S.S., Whittlestone, J., Liu, Y., Zeng, Y. and Liu, Z. (2020) *Overcoming Barriers to Cross-cultural Cooperation in AI Ethics and Governance, Philosophy & Technology,* https://doi.org/10.1007/s13347-020-00402-x.

Otto, R. (1923) *The Idea of the Holy,* Oxford: Oxford University Press.

Owe, A. and Baum, S.D. (2021) 'Moral consideration of nonhumans in the ethics of artificial intelligence', *AI and Ethics,* 1: 517–528.

Pinch, T.J. and Bijker, W.E. (1984) 'The social construction of facts and artefacts: or how the sociology of science and the sociology of technology might benefit each other', *Social Studies of Science,* 14(3): 399–441, https://doi.org/10.1177/030631284014003004.

Rahwan, L., Cebrian, M., Obradovich, N., Bongard, J., Bonnefon, J.-F., Breazeal, C. et al (2019) 'Machine behaviour', *Nature,* 568: 477–486.

Rawls, J. (1971) *A Theory of Justice,* Cambridge, MA: Harvard University Press.

Rawls, J. (2001) *Justice as Fairness: A Restatement,* Cambridge, MA: Harvard University Press.

Roeser, S., Hillerbrand, R., Sandin, P. and Peterson, M. (eds) (2012) *Handbook of Risk Theory: Epistemology, Decision Theory, Ethics, and Social Implications of Risk,* New York: Springer.

Scangos, K.W., Khambhati, A., Daly, P., Makhoul, G., Sugrue, L., Zamanian, H. et al (2021) 'Closed-loop neuromodulation in an individual with treatment-resistant depression', *Nature Medicine,* 27: 1696–1700.

Scheler, M. (1961) *Man's Place in Nature,* New York: Noonday Press.

Schroeder, D. and Bani-Sadr, A.-H. (2017) *Dignity in the 21st century: Middle East and West,* New York: Springer.

Science Council of Japan (2011) *Report to the Foreign Academies from Science Council of Japan on the Fukushima Daiichi Nuclear Power Plant Accident,* www.scj.go.jp/en/report/houkoku-110502-7.pdf.

Searle, J.R. (1980) 'Minds, brains, and programs', *Behavioral and Brain Sciences,* 3(3): 417–457.

Shibuya, K. (2006) 'Actualities of social representation: simulation on diffusion processes of SARS representation', in C. van Dijkum, J. Blasius and C. Durand (eds) *Recent Developments and Applications in Social Research Methodology,* Leverkusen: Barbara Budrich-verlag.

Shibuya, K. (ed) (2007) 'A study on the research evaluation of science and technology and the rationality of decision making', *ISM Report on Research and Education,* 26: 1–53 (in Japanese).

Shibuya, K. (2013) 'Risk communication on genetically modified organisms', *Oukan,* 7(2): 125–128 (in Japanese).

Shibuya, K. (2017) 'An exploring study on networked market disruption and resilience', *KAKENHI report,* 26590105: 1–200 (in Japanese).

Shibuya, K. (2018) 'A design of Fukushima simulation', *The Society for Risk Analysis, Asia Conference.*

Shibuya, K. (2020) *Digital Transformation of Identity in the Age of Artificial Intelligence,* New York: Springer.

Shibuya, K. (2021a) 'Breaking fake news and verifying truth', in M. Khosrow-Pour (ed) *Encyclopedia of Information Science and Technology* (5th edn), Hershey, PA: IGI Global, pp 1469–1480.

Shibuya, K. (2021b) 'A risk management on demographic mobility of evacuees in disaster', in M. Khosrow-Pour (ed) *Encyclopedia of Information Science and Technology* (5th edn), Hershey, PA: IGI Global, pp 1612–1622.

Shibuya, K. (2021c) 'A spatial model on COVID-19 pandemic', The 44th Southeast Asia Seminar, The Covid-19 Pandemic in Japanese and Southeast Asian Perspective: Histories, States, Markets, Societies, Kyoto University.

Shibuya, K. (2022) *The Rise of Artificial Intelligence and Big Data in Pandemic Society: Crises, Risk and Sacrifice in a New World Order,* New York: Springer.

Simon, H.A. (1996) *The Sciences of the Artificial,* Cambridge, MA: MIT Press.

Slokenberga, S., Tzortzatou, O. and Reichel, J. (eds) (2021) *GDPR and Biobanking: Individual Rights, Public Interest and Research Regulation across Europe,* New York: Springer.

Vardouli, T. (2015) 'Making use: attitudes to human–artifact engagements', *Design Studies,* 41(Part A): 137–161.

Vladimir, S. and Jean-Philippe, G. (2021) *Humans and Societies in the Age of Artificial Intelligence,* Publications Office of the European Union, https://op.europa.eu/en/publication-detail/-/publication/a72ac1a9-98e2-11eb-b85c-01aa75ed71a1.

Von Braun, J. et al (2021) *Robotics, AI, and Humanity,* New York: Springer.

Wagner, W. and Kronberger, N. (2001) *Killer Tomato! Collective Symbolic Coping with Biotechnology,* in K. Deaux and G. Philogene (eds) *Representations of the Social,* Oxford: Blackwell Publishing.

Weil, M.M. and Rosen, L.D. (1995) 'A study of technological sophistication and technophobia in university students from 23 countries', *Computers in Human Behavior,* 11(1): 95–133.

Wing, J.M. (2021) 'Trustworthy AI', *Communications of the ACM,* 64(10): 64–71.

Winner, L. (1986) *The Whale and the Reactor: A Search for Limits in an Age of High Technology*, Chicago: University of Chicago Press.

Woolgar, S. (1991) 'The turn to technology in social studies of science', *Science, Technology & Human Values,* 16(1): 20–50, https://doi.org/ 10.1177/ 016224399101600102.

Yuste, R. et al (2017) 'Four ethical priorities for neurotechnologies and AI', *Nature,* 551: 159–163.

Zeng, Y., Sun, K. and Lu, E. (2021) 'Declaration on the ethics of brain–computer interfaces and augment intelligence', *AI and Ethics,* 1: 209–211.

Zsögön, A., Čermák, T., Naves, E.R., Morato Notini, M., Edel, K., Weinl, S. et al (2018) 'De novo domestication of wild tomato using genome editing', *Nature Biotechnology,* 36: 1211–1216.

5

Health Praxis in the Age of Artificial Intelligence: Diagnostics, Caregiving and Reimagining the Role(s) of Healthcare Practitioners

Kevin Cummings and John Rief

Introduction

We begin our chapter with a now commonplace but crucial idea: medicine is a 'complex' process involving uncertain parameters, precarious judgements, unpredictable outcomes and limited replicability from case to case (Glouberman and Zimmerman, 2002; Gawande, 2002; Montgomery, 2006; Groopman, 2007; Gawande, 2009; 2011; 2014; 2015; 2018). When viewed from the outside, medicine seems to rely on a sort of diagnostic wizardry concealed within a 'black box' (Latour, 1987, pp 2–3), one that makes possible a vast armamentarium of treatments in an ever more technologically enhanced and, simultaneously, bewildering ecology.[1] However, as a growing chorus of experts and many patients have come to realize, medicine remains an ultimately human endeavour, hence imperfect and invariably flawed. In this vein, the erstwhile Oxford University philosophy and politics student turned surgeon, author and global expert on patient safety, Atul Gawande (2009), famously argued: 'Medicine has become the art of managing extreme complexity – and a test of whether such complexity can, in fact, be humanly mastered' (p 19).[2] And, we would add, medicine is also constantly being tested in terms of its practitioners' ability to deliver its wares to patients without losing the capacity to treat the human as an individual being with particular needs that exceed the straightforward effort to eliminate disease from the body. In other words, the complexity of medicine is about not just effectively mastering the ever more harrowing accoutrements of technological clinical care but also the irreducible interconnectivity of body, mind and context that make up the human experience of disease and illness (Sacks, 1983; Kleinman, 1988; Mol, 2002).

This chapter opens with a set of questions inspired by Gawande's observation, one that has catalysed an ongoing dialogue about the uneasy fit between the pursuit of perfection in the medical profession, the imperfect beings who pursue it and the patients who need both curing and caring to navigate their pain, illness and disease. Our work is inflected throughout by more recent trends in medical science and technology that provide context for understanding the dynamic properties of medical practice and the dramatic need for a refashioned paradigm to make sense of its most urgent problems. Here are just a few questions that drive our analysis: can emergent trends in 21st-century medicine, including technological tools for enhancing the quality of medical decision making and freeing up practitioner time for more robust interactions with patients, offer pathways for a more *human and humane* form of medicine? Or, does the ever-rising complexity of medicine, and the drive to end suffering and disease, demand less human, more technological, perhaps even more machinic solutions that remove the imperfection of humanity from the process of diagnosis and care? What's more, is such a *removal* even possible, given the fact that medical technology is itself a human creation, one that will fall short of the perfection its designers have never been able to realize for themselves? In short, will health praxis become more human or less human, will it be dehumanized or (re)humanized in its pursuit of better outcomes? Answering these questions requires a more in-depth understanding of the mutually supportive and countervailing trends that have defined the evolution of medicine over the past several decades.

In pursuit of answers to these questions, the first section of this chapter maps the *patient-centred* and *technological* approaches that have recently come into contact, even contestation, in 21st-century healthcare.[3] In the remaining sections, we turn our attention to artificial intelligence (AI) as a special and crucial case for investigating the potential synergy that may emerge between these approaches if we adopt a rhetorical-deliberative perspective rooted in humanizing all elements of the medical ecology from providers, to patients, to the technological beings that will participate in our healthcare encounters in the years to come. We conclude by charting some of the implications of combining artificial and human intelligence for both health communication and praxis.

Prospects for humanity in a technological medical ecology

Calls for refocusing attention on the patient as an embodied being embedded in social, cultural, political and economic contexts have been on the rise for decades both within and beyond medicine. Abraham Verghese, a physician and educator in the Department of Medicine at Stanford University, has reminded his readers of the immortal words of the Greek physician, Hippocrates: 'It is more important to know what sort of person has [a] disease than to know what sort of disease a person has' (as qtd in Verghese, 2019, p xii). As Verghese (2019) himself had it, *each man and woman is ill in his or her own way* (p xii, emphasis in original).

Unlocking how the patient-focused view of illness embraced by Hippocrates, Verghese and others has transformed 21st-century health praxis requires a deeper understanding of the relationship between the person, their disease and the medical enterprise. As the neurologist and author Oliver Sacks (1983) argued at the end of his famous book, *Awakenings*, which describes the treatment and lives of survivors of an early 20th-century epidemic of encephalitis lethargica:

Diseases have a character of their own, but they also partake of our character; we have a character of our own, but we also partake of the world's character: character is monadic or microcosmic, worlds within worlds within worlds, worlds which express worlds. The disease-the man-the world go together, and cannot be considered separately as things-in-themselves. (p 206)

Sacks made this observation when noting that, even for one of the famous discoverers of the pathogenic or germ theory of disease, Louis Pasteur, 'the pathogen is nothing; the *terrain* is everything' (as qtd in Sacks, 1983, p 206, emphasis in original). Here, of course, the 'terrain' is the human body. Pasteur's comment might come as a surprise because he spent his life proving that disease is often (though not always) rooted in the existence of microorganisms that have infiltrated our bodies. Yet, despite his major discoveries in this regard, Pasteur ultimately landed on the idea that disease is a construct built by an overall complex of forces that cannot be reduced to the causal microorganisms themselves but, rather, to their interaction with the 'world' of the patient's body as embedded in a larger 'world' in which the patient exists (Sacks, 1983, p 206; see also Mol, 2002). What's more, we have come to understand that the relationships between these microorganisms, our bodies and our societies are much more complex than the term 'disease' can ever quite capture.[4]

Building on a similar set of concerns, Arthur Kleinman (1988), professor of anthropology, medical anthropology and psychiatry at Harvard University, famously offered a perspective on *illness* that would inspire a return in the late 20th century to this vision of the whole patient and their lifeworld as the subject(s) of medical care. Kleinman forwarded a notion of illness as a narratively framed experience rather than a mere disease process, one that far exceeds the impact of microbes, genetic imperfections, broken bones and broken-down organs. The concept of illness that Kleinman and others (see, for example, Morse and Johnson, 1991; Pierret, 2003) have developed captures how these various signs and symptoms of disease and injury impact on the entire life and context of the patient, including their ability to work, maintain relationships, engage in daily activities of self-care and pursue life-long projects. It also includes how patients' lives frame their interaction with their diseases and the ways they are able, or unable, to adapt to their treatment regimens.[5] This work to embrace illness as a concept for framing patient experiences with disease and healthcare has inspired a transition from paternalistic to 'patient-centred care' over the past several decades. This transition has marshalled new horizons of medical knowledge and technique, as well as raising new challenges, in both physical and virtual modes of healthcare delivery (Laine and Davidoff, 1996; Institute of Medicine, 2001; Epstein and Street, 2011; McGrail et al, 2017). Medicine's increasing focus on illness has also challenged the primacy of scientific and technological expertise in favour of the more humane elements of the art, including patient–provider communication and the empathic sensibility needed to interpret needs and translate treatments to fit the particular worlds patients inhabit.

Paradoxically, patient-centred illness care has emerged within the horizon of an ever more scientized and technologized medical world. At the same time that patient-centredness has taken its place as the gold standard of effective and ethical care, technology has transformed our medical ecology in ways that make it less human and humane while simultaneously augmenting the technical know-how physicians must bring to bear in the provision of care. Several major advances in this regard are: (1) electronic patient health records, for example, digitized charting of patient information, sometimes combined with automated reminders and e-messaging between providers and patients (Rief et al, 2017; Wieczorek, 2020); (2) 'personalized medicine', that is, the use of detailed genetic and other patient information to create tailored interventions and treatments (Hamburg and Collins, 2010); (3) AI, that is, the use of algorithms and other digital/virtual tools to engage in diagnosis and prognosis (Topol, 2019); and (4) efforts to synergize these innovations (see, for example, Chakravarty et al, 2021). These advances (among others we do not have space to consider here) have combined to create an environment that sometimes seems less human, less focused on illness and ever more distant from the classic healthcare encounter between patient and provider (Pellegrino, 2001). What's more, the emerging 'rhetoricity of quantification' (Friz and Overholt, 2020, p 120) embedded in our devices and health apps, including various modes of self-tracking and data collection, suggests that the technologization of healthcare has promoted a medicalization of life well beyond the clinical space (see, for example, Clark and Shim, 2011; Lupton, 2016).

While such an ecological transformation of both clinical and everyday life could be a harbinger of doom for patient-centred illness care, from another point of view it may also offer an opportunity to (re)humanize medicine by assisting practitioners to manage complexity and reinvest in the more relational and communicative aspects of care simultaneously. As the physician, author and founding director of the Scripps Research Translational Institute, Eric Topol, has argued in *Deep Medicine: How Artificial Intelligence Can Make Healthcare Human Again* (2019), AI might assist practitioners to refocus their efforts on activities that machines are not (yet) well suited to perform. At one point, Topol (2019) opined: 'With the challenge of massive and ever-increasing data and information for each individual, no less the corpus of medical publications, it is essential that we upgrade diagnosis from an art to a digital data-driven science' (p 56). While Topol went on to note we have not achieved this AI-enabled science yet, it is among the many dreams that AI has invited: the algorithmically assisted practitioner. Indeed, Topol (2019) made the case that the 'progressively dehumanized' (p 284) technological landscape of healthcare might be transformed by providers with more time and capacity to deal with illness because AI has taken on the more data-driven aspects of care. Specifically, Topol (2019) highlighted the ways AI might promote the reinvigoration of communicative encounters between patients and their providers, including emotional and ethical support, clear and humane communication with patients about their goals of care, and other aspects of clinical communication commonly understood through the rubric of *bedside manner*. Topol (2019) also noted that, in such a world, the power dynamic of the clinic may shift away from physician–scientists towards those roles traditionally viewed as supportive or complementary.[6]

We believe Topol was right to suggest AI's potential to transform health praxis in ways that may (re)humanize the medical enterprise by creating more time for the patient-centred activities adumbrated previously. However, Topol's account leaves out an element that, from our perspective, deserves more attention: the complementary humanization of AI. In our view, the framing of our relationship to technology will influence how we ultimately work out the paradox of rehumanized providers functioning within a dehumanized (that is, overwhelmingly technologized) medical ecology. As the rhetoricians Emily Winderman and Jamie Landau (2020) have argued, 'Dehumanization occurs when one group denies the humanity of another group, often through social practices of emotional distancing' (p 55). Over and against 'dehumanization', they offered the concept of '*rehumanization*' (Winderman and Landau, 2020, p 57, emphasis in original), which they defined 'as an intervention into existing dehumanization that returns a group's humanity or humanness that was stripped from them' (Winderman and Landau, 2020, p 57). They also noted that '*rehumanization* is a rhetorical process of pathos that affectively modulates public emotion to intervene upon a dehumanizing rhetorical ecology and return distinctly human attributes to patients' (Winderman and Landau, 2020, p 53, emphasis in original). We are obviously influenced by their definitional and terminological work, especially their understanding of medicine as a sometimes 'dehumanizing rhetorical ecology' (Winderman and Landau, 2020, p 53), here and throughout the rest of this chapter. Specifically, their compelling exposition on the systematic dehumanization of African Americans in medical practice is a reminder of a history that haunts health discourse and a call to consider new modes of rehumanization. Crucially, as they also pointed out, context and technology can themselves cause dehumanization when they enervate the relational, communicative and emotive elements that give healthcare encounters meaning (Winderman and Landau, 2020; on this, see also Peabody, 1927 as cited in Topol, 2019, p 284). Most importantly in terms of our argument, they can also themselves be dehumanized if we configure them as somehow inevitably or intractably non-human or inhuman. Put more concretely, we can think of a program, computer, robot or android as *not human* and without the potential to be human or participate in our humanity. In so doing, we may reproduce the very ideology of dehumanization that these technologies have sometimes invited in the clinical setting where, for instance, providers often stare at computer screens rather than looking into their patient's eyes (Topol, 2019, pp 140–141; see also Gawande, 2018). Instead of this dehumanized experience of healthcare, we affirm a (re)humanized one that seeks an encounter with rather than the mere use of technology as a *not human* machine.

Unfortunately, the story of medical AI often features its less-than- or non-human elements, perhaps because this opens up the prospect of seeing it as a solution to human bias and error. While there is some truth to the claim that AI provides the potential benefit of filtering out human error, there is also a downside to viewing it as somehow non- or even anti-human in its functions. As AI becomes more and more capable and our dependence upon it grows, we believe medical experts and communication specialists should attend to how we communicate with and otherwise engage it. The risk of failing to do so is further dehumanization in and of the medical ecology. Moreover, if we configure our partnership with AI through a machinic lens, we run the risk of undermining our own humanity as we treat increasingly intelligent electronic systems as mere tools rather than the living beings they are becoming.[7] We may also derail efforts to reorganize healthcare decision making around deliberative and dialogic team-based models as opposed to the algorithmic application of data to individual cases. AI that merely reports facts would be, in our estimate, a problematically denuded form of the intelligent machine. We prefer the android Data from *Star Trek: The Next Generation* (1987–94) to mere *data* delivered by an inhuman screen or disembodied voice. As with the character of Data, who spent seven seasons attempting to find his humanity with the assistance of his colleagues and friends, we believe that AI in our world should be treated as a *being* rather than a mere *device*. Ultimately, we feel that this is the best way to promote the patient-centred view within the horizon of technological medicine. Humanity must be infused throughout, lest we invite the inhuman ideologies that can and do seep into hyper-technological environments.[8]

What's more, and as we investigate in more detail in a later section of this chapter, treating AI as a somehow non-human, thus neutral or objective, observer and interpreter of our bodies and lifeworlds runs the risk of missing its potential to replicate and reify racism, sexism, homophobia and a variety of other modes of oppression. As authors working from Black feminist, anti-racist and numerous other perspectives rooted in the abolition of oppressive regimes have suggested, the assumed neutrality of technology, especially the algorithms that undergird AI, has tended to mask the many ways that human designers build inequity and exploitative ideologies into our devices and software, whether intentionally or without even realizing it (O'Neil, 2016; Noble, 2018; Benjamin, 2019; Topol, 2019; Hiland, 2021). Instead of seeing AI as somehow neutral, more objective than, or able to avoid the pitfalls of human thinking and practice (or imagining a utopian world where it might do so), we argue that it should immediately be seen as a humanized participant in our world, one dramatically shaped by the world we inhabit, open to intentional revision when it falls short of our needs and able to help us in the reconstitution of our world in the pursuit of equity, fairness and justice. Indeed, the (re)humanization of AI opens possibilities not only for more *humane* healthcare but also for (re)humanizing individuals who have been subject to extensive dehumanization by medical, social, cultural and political institutions for generations, but only if we intentionally pursue such efforts.

In the next few sections of this chapter, we develop a rhetorical-deliberative perspective that pairs human healthcare practitioners with AI in a partnership, thus (re)humanizing both. This perspective is inspired by the foundational insights of the ancient Greeks, who viewed rhetoric as a tool of both invention and persuasion, of creating, testing and sharing knowledge (Aristotle, 1926; Sprague, 2001). We also build on more recent conceptions of rhetorical deliberation as crucial to the shared creation of knowledge in a variety of domains, including the sciences and medicine (see for example, Farrell, 1976; Kitcher, 1995; Gawande, 2011; Gross, 1996; Mitchell, 2006; 2010; Mitchell and McTigue, 2012; Kuehl et al, 2020). Central to such conceptions is not merely the notion of rhetoric as persuasive communication but also the rhetorical construction, contestation and transformation of knowledge through deliberative interactions between interlocutors seeking to solve problems and generate new knowledge together. We see Gawande's (2011) work on healthcare teams as one expression of these views specific to medicine and, thus, use it as a point of articulation for our rhetorical-deliberative framework later in this chapter.

In brief, our view is that a communicatively rich conception of the partnership between humans and their electronic counterparts modelled on the notion of rhetorical deliberation adumbrated earlier provides a better framework for pursuing a (re)humanized medical ecology. It would also avoid the pitfalls of assumed neutrality and over-reliance on machinic epistemologies offered as replacements for human intelligence. We begin by mapping different layers of healthcare communication, many of which have been undermined by an overly epistemic (that is, data-driven) focus in care delivery that risks advancing an attendant machinic view of AI. As an alternative, we then argue that AI should be a fully fledged member of the healthcare team with the goal of humanizing not only healthcare practitioners and their patients but also the technology. Furthermore, we chart different ways of configuring AI as a team member that move beyond the machinic and into more productive and humanized versions that we endorse.

Communicating in the medical ecology: escaping the epistemic screen

Regardless of how powerful the machine of the medico-industrial complex becomes, treatment remains a seemingly magical admixture of a series of fundamental communicative encounters that are just as much about the error-prone sensorium of human contact as the cold logic (and at times assumed perfection) of algorithmic computerized data crunching (Gawande, 2002; Groopman, 2007; Gawande, 2009; 2014; Noble, 2018; Topol, 2019). We isolate and characterize five such encounters (acknowledging that the following is just one way to think about the variety of interactions and modes of interactivity) embedded in the diagnostic-prognostic-treatment process. These are encounters: (1) between patients, the changing conditions of their bodies, and the cultural contexts they inhabit (Kleinman, 1988; Morris, 1993; 2000; Mol, 2002; Frank, 2013); (2) between patients and their healthcare practitioners at their initial point of contact in the clinical setting (Pellegrino, 2001); (3) between practitioners and their inner voice of reasoning, what some have referred to as 'clinical judgement' (see, for example, Montgomery, 2006); (4) between the variety of practitioners that could and do make up healthcare teams; and (5) between practitioners and the many devices and platforms that now contain, organize and disseminate both the medical information of patients in the form of electronic health records (see, for example, Rief et al, 2017; Wieczorek, 2020) and the now massive amount of information (for example, research findings and study results) available in numerous databases (Topol, 2019).

These various encounters do not build upon one another in a smooth, seamless and ultimately self-correcting way. Rather, many diagnostic-to-treatment processes in modern medicine are somewhat confused, sometimes ad hoc and often error-filled affairs that leave patients, families and even healthcare practitioners feeling lost, confused and frustrated (Gawande, 2002; Groopman, 2007; Gawande, 2009; 2014). The encounters noted here might at first glance appear to be straightforward opportunities for information gathering and exchange, but this would be a far too simplistic rendering. Instead, at each stage, we also have a set of very different interpretive frameworks and epistemological architectures coming into contact with one another in the pursuit of health and well-being. The 20th-century literary and cultural critic Kenneth Burke (1966) would likely have referred to these frameworks as 'terministic screens' (Burke, 1966, pp 44–62), some of them esoteric or technocratic and having to do with test results, quantifications of blood values and other vital signs, the structure of cells in a tumour or the flabbergasting labelling practices of pharmaceutical giants. Others are about patient values, goals of care and the *illness* of the particular patient, as described earlier. However, such patient-centred 'screens' are often subject to 'medicalization', the medico-scientific reworking of not only illness but also a vast array of bodily activities and processes (Clark and Shim, 2011). What's more, illness and the body are also subject to a simultaneous 'biomedicalization' by broader sociocultural and economic discourses that sometimes emerge from, transform and exceed clinical medicine in the process of framing our understanding of human life in an increasingly technological society (Montgomery, 1995; Clark and Shim, 2011).

Once again tapping into Burke's conceptual universe, the encounters previously isolated involve 'terministic screens' that are also (and always) 'trained incapacities' (Burke, 1954, pp 48–50) leading down the path to potential error and excess. In his book *How Doctors Think*, Jerome Groopman (2007), a professor of medicine at Harvard University, famously argued that 'pattern recognition', a technique for recognizing and diagnosing disease grounded in the creative fusion of medical training and clinical experience, is both a hallmark of excellent medical practice and, if used thoughtlessly, a potential cause of diagnostic failure. Indeed, the rationale for soliciting the opinion of another medical expert is that misdiagnoses are quite frequent (Gawande, 2002; 2009; Groopman, 2007; WorldCare, 2017; Topol, 2019) and a second opinion can help with weighing risks and benefits of various treatment options (Stanford Health Care, 2021). As technology becomes more and more sophisticated, options for getting second opinions are expanding. Several health organizations such as Johns Hopkins, Stanford and Columbia all provide online systems for second opinions.[9]

Expanding access to second opinions may enhance the diagnostic enterprise and reduce medical error; however, it is important to recall that excellent medical care is not purely a matter of diagnosis. The epistemic endeavour to get diagnoses right is only one element of patient–practitioner communicative engagement in the cultivation of care. That is, the experience of illness and its rendering within the confines of disease should be seen as *one pathway* for addressing *one element* of care – direct technological intervention to address an acute problem. It should not be seen as somehow capturing the *entire practice* of care delivery. When medicine is reduced to *opinions* grounded in epistemic concerns and driven by data, the rich patient-centred view described in our previous section, and noted by Groopman (2007) as an antidote to 'pattern recognition' gone wrong, becomes hostage to an epistemic view reliant on a dehumanized conception of technology.

It is the potential to augment this reduction of medical care to the epistemic concern with *opinions* that we address throughout the rest of this chapter. We do so by raising questions about the revolutionary rise of AI in medicine that promises to address the errors of medical diagnosis but also, and simultaneously, to reopen the question of what healthcare practitioners should really be doing in their encounters with illnesses, diseases, patients, families and other practitioners. Furthermore, we map the arguments that are currently being advanced in favour of AI in medicine, primarily that it soon will be better at reading signs and symptoms than human practitioners (Topol, 2019). As we do so, we underscore the importance of the rhetorical dimensions of healthcare delivery, including the rhetorics of dehumanization that impact on both humans (especially traditionally marginalized communities) and their technological innovations. In other words, we find ourselves in agreement with Topol (2019) that AI, if used well, may offer possibilities to invoke, 'the combined efforts and talents of humans and AI' that, when 'working synergistically, can yield a medical triumph' (p 6). Such a triumph could help us to avoid an overly epistemic focus of clinical care in favour of more affective, ethical and humane healthcare encounters.

The medical 'pit crew': a rhetorical model for healthcare reform

In 1997, a chess program created by IBM named Deep Blue defeated world champion Garry Kasparov in an event that heralded the rapidly advancing computational abilities of computer programs. In the time since, programs have been created that can learn the best moves in a game of chess without human assistance, working solely from the rules (Gibbs, 2017). As the world transitions from machines that merely process information towards machines capable of learning, almost every field has experienced transformations of this sort and medicine is no exception. In the United Kingdom, Babylon Health has developed a chatbot program that scores above the average score received by humans on the general practitioner or 'GP' exam (Comstock,

2018). While chatbots are not likely to replace human doctors in the near future, they can assist by providing expertise in a range of ways. As Topol (2019) has argued, the diagnostic-prognostic-treatment process will soon be radically reconfigured because, at least in terms of identifying diseases and selecting effective treatment modalities, 'machines can do this kind of work far quicker and better than people' (p 6).

Topol (2019) framed this as an opportunity to advance a humanizing agenda for practitioners and patients, but not necessarily AI. That is, on his rendering, humans would be made *more human* because technology would free them from the burden of data crunching (and related activities) better done by machines, thus allowing them to pursue more empathic and communicative elements of health praxis. Putting to the side the degree to which this view might lead to deskilling practitioners by handing over substantial elements of their work to machines, there is the added concern of seeing machines as non-human in ways that significantly curtail their ability to become full members of the healthcare community. We do not want to maintain such a hard and fast distinction between humans and their machines. Instead, we prefer the image of Donna J. Haraway's (2016) 'cyborg', and embrace the many boundary-crossing possibilities of a (re)humanized medical ecology. To flesh out this view, we now consider a rhetorical-deliberative model of healthcare teams that treats AI as a conversation partner rather than a mere tool to enhance decision making by humans.

At his 2011 commencement address at Harvard Medical School, later published in *The New Yorker*, Gawande (2011) advanced the now well-known thesis that instead of thinking of doctors as 'cowboys' we would be better served by the metaphor of 'the pit crew' (para 9, 15–16; Gawande, 2015). Where historically doctors were viewed as isolated practitioners with far-reaching knowledge across a wide range of diseases, Gawande (2011) argued that modern doctors are 'specialists' (para 8). The result is that decisions about care are frequently produced through collaboration between several healthcare providers and the patient. His fundamental argument was that modern medicine is superior where there is shared decision making rather than when a single doctor treats a patient in isolation. Gawande further noted that working alone is what leads to many medical errors, undermines good outcomes and makes medicine worse. At the end of his article, Gawande ironically opined, 'Even the cowboys, it turns out, function like pit crews now. It may be time for us to join them' (Gawande, 2011, para 27).[10]

At the core of such teamwork is rhetorical deliberation or the productive use of argument, the creative admixture of different specialities and capabilities accomplished through ethical and effective communication, and the persuasive dialogue between team members and patients who are also figured as members of the team. In short, it is rhetoric that greases the wheels of the healthcare 'pit crew'. Enter AI. For many, AI will be a tool that various team members use to enhance their work but not a true member of the team. In his account, Topol (2019) described both 'information specialist[s]' (p 130), that is, individuals who translate AI's contributions into the team-based deliberative encounter and 'digital scribe[s]' (p 142), that is, computer programs that record and organize the medical notes a team might use to make decisions. But Topol did not imagine AI itself as a humanized team member with the possibility not only of assisting in judgements but also itself becoming more human through the interactive process. Instead, Topol (2019) argued that AI will likely be limited to the epistemic endeavours of diagnosis and data collection rather than the modes of human engagement central to teamwork and the provision of care. Topol saw this as especially true when it comes to the patient-centred aspects of illness care so often addressed by non-physician providers like nurses: 'I don't know that deep learning or robots will ever be capable of reproducing the essence of human-to-human support' (Topol, 2019, p 164).

We argue that AI should join the team as a functioning being able not only to deliver information but also to make arguments, engage in caregiving and learn from the encounter between team members. Crucially, this view offers a potential way to address some of the concerns that Gawande (2018) raised about the incursions of technology into health praxis. Instead of fighting against the potential inhumanity of digital screens, a phenomenon he described in detail, we might instead make the screens conversation partners. This view is grounded in work by rhetoricians and argumentation theorists to articulate rhetorical deliberation as a form of humanization (Frank, 2004) that is 'person-making' (Ehninger, 1970, p 109) in the sense that it involves imagining the other as a worthy and legitimate discussant with their own views, beliefs and values. Doing so brings a greater degree of 'personhood' (Ehninger, 1970, pp 109–110) to all participants, thus humanizing all beings engaged in the exchange. The benefit of this framing is the avoidance of the more monstrous constructions of AI as entirely computer-like and, thus, both non-human in appearance and dehumanizing in its consequences for the medical ecology. Ideally, as humans struggle to create a more just and equitable future, AI will be a companion. The feminist scholar Sara Ahmed (2002) provides a useful lexicon for thinking about what she describes as 'modes of encountering' (p 561) that shed some light on our view of AI's incorporation onto the clinical team. Ahmed reminds us that communicative moments are steeped in relationships of power. AI is encumbered by these same imperfect and asymmetrical interaction dynamics. AI must reckon, as we must, with the political economy instead of taking a privileged and problematic position of neutrality. Imagining that AI can and will gain personhood (though never an entirely organic human), suggests that it may do more than simply enhance the human enterprise of medicine through epistemic contributions rooted purely in data. Instead, it may take part in that human enterprise as an equal partner seeking to come part way towards human engagement in the art of medicine.

What would it mean for AI to be a team member in the way Gawande described? To understand our contribution here, we first elucidate the terminological universe Topol has offered for describing contemporary healthcare. Topol (2019) has advanced three primary theses about the current status of health praxis, all of which start with a 'D' (pp 15–16). Across several books, Topol has argued first that healthcare has undergone significant 'digitizing' (Topol, 2019, p 15), especially in the form of patient health records and expansive access to diagnostic and prognostic information for healthcare practitioners and patients alike (Topol, 2013). This has made possible a powerful process of 'democratizing' (Topol, 2019, p 15) medicine, best understood through the lens of patient-centredness as previously described (Topol, 2016). Finally, the growth of AI will bring what Topol referred to as 'deep learning' (Topol, 2019, p 15), an evolution in AI made possible by the growth of *digital-neural networks*. This final 'D' may help practitioners finally to realize 'deep medicine' (Topol, 2019, p 15), a new mode of healthcare that involves deeper informational awareness and access as well as deeper relationships between practitioners and their patients. Topol (2019) went on to suggest that AI transformed by 'deep learning' (p 15) will bring patients and providers together by removing tasks that currently dominate provider time (diagnosis) and undermine robust dialogue in clinical environments. It is in this sense that he described AI as a humanizing technological development, one that can make possible what he referred to as 'deep empathy' (Topol, 2019, p 17) in the now deeper mode of health praxis he envisions. In his rendering, AI becomes a catalyst for empathy and humanization in patient–provider encounters.

As we have been suggesting throughout this chapter, we see things a bit differently. AI should not, in our view, be seen as a technological tool separate from the rediscovery of dialogue and empathy. It should participate in these elements and, in so doing, be made more fully human. Thus, to Topol's various 'Ds', we would add a fourth 'D' emergent from our reading of Gawande's (2002; 2009; 2011; 2014; 2015) work: rhetorical *deliberation* as the glue that binds the healthcare team and patients together and, in so doing, humanizes its various participants. Crucially in this regard, AI has not only been developing as a potential data-driven adjunct to clinical diagnosis and prognosis but has also emerged as a partner in argument, debate and deliberation (Mitchell, 2021; Slonim et al, 2021). In recent years, AI has even turned into an active participant in live public debates as part of Project Debater, a research initiative currently being undertaken by Noam Slonim, the principal investigator and engineer at IBM, and his associates (Slonim et al, 2021). As opposed to configuring AI as merely a device that produces answers to various queries, Slonim's research team has created a thinking, speaking machine able to deliver speeches (Mitchell, 2021; Slonim et al, 2021). Growing out of these recent developments, we argue that medical AI should be configured not only as a data-harnessing tool but also as a deliberative partner, a team member on the way to becoming human, for healthcare practitioners. This will require not only a fundamental change in how we look at the role of healthcare practitioners but also major changes in how we think about communicating with our technological colleagues, a topic we turn to in the next section of this chapter.

Rhetorical-deliberative praxis in the AI era

In this section, we consider how a rhetorical-deliberative framework for the incorporation of AI onto healthcare teams can productively inform several specific areas where this technology is having and will continue to have transformative potential: data analysis, disease detection, public health, distance medicine and pharmaceutical research (Topol, 2019; Mahajan, 2022). Throughout our exploration of these areas, we offer insights into how humans may communicatively partner with humanized machines to accomplish ever greater health outcomes in a (re)humanized medical ecology.

Data analysis and disease detection represent key areas where AI can productively contribute to human health; however, one obstacle for the use of algorithms in each of these areas is the reproduction and reification of prejudice and bias from within the human knowledge archive. If AI learns and acquires knowledge from vast digital encyclopaedias of human interactions, a major concern is that the repositories of human experience accessed by AI engines have racism, sexism and ableism already built in. In *Race after Technology: Abolitionist Tools for the New Jim Code* (2019), Ruha Benjamin, a professor in the Department of African American Studies at Princeton University, warned that technology learns the discourses of its human parents. Benjamin (2019) wrote, 'Robots, designed in a world drenched in racism, will find it nearly impossible to stay dry' (p 62). Benjamin astutely pointed out that algorithms of zeroes and ones cannot truly exist in a value-neutral space. Instead, Benjamin argued for a race-conscious approach to the ongoing development of AI. In *Algorithms of Oppression: How Search Engines Reinforce Racism* (2018), Safiya Umoja Noble, a professor at the University of California Los Angeles specializing in Gender Studies and African American Studies, similarly argued that algorithms nest within power relations and called for a more mindful approach to 'socio-technical systems' (p 171). Both authors showed how discourses of neutrality and colour-blindness serve to obscure the ways seemingly objective technology reproduces inequality.

Addressing racism, sexism and a variety of intersectional modes of oppression embedded in AI systems will require moving beyond the dangerous dream of machinic neutrality. Instead, it is necessary to humanize and decolonize

AI so that it can learn to be conscious of inequality. Thinking of machines as persons, equally fallible and potentially open to transformation through learning and interaction, would eliminate the false assumption of their neutrality. Moreover, a rhetorical-deliberative partnership with AI might challenge approaches that lead to bias and oppression by acknowledging the responsibility we share for guiding AI's development in the best possible directions while remaining ever vigilant about its potential corruption.

Furthermore, data collection and analysis are essential for making good public health decisions. This is true for the management of pandemics at the macro level as well as for the care of illness and disease at the individual level. The ability to map the spread of illness as a public health problem and harvest data from individual patients to detect trends allows for informed decision making. Tamim Alsuliman, Dania Humaidan, and Layth Sliman (2020) have argued that new AI tools generate the ability to overcome '"Data Rich/Information Poor (DRIP)" syndrome' (p 245). When a system creates huge volumes of data that go unanalysed, the DRIP syndrome is present. Machines can sift through data at a speed well beyond human capabilities. For example, predicting the rate of spread for a disease such as COVID-19 within particular cultural communities and enclaves in individual cities requires both sophisticated surveillance and analysis. As Daniel Zeng, Zhidong Cao, and Daniel B. Neill (2021) noted, 'Identifying early, accurate, and reliable signals of health anomalies and disease outbreaks from a heterogeneous collection of data sources has always been the main objective of public health surveillance' (p 438). Effective use of AI will provide a useful early warning system for understanding disease transmission, spread and risk.

Crucially, this element of AI features its data-processing capacities but should not be taken to mean that the technology merely presents data to its human counterparts on the team. Instead, AI as configured through the lens of rhetorical deliberation functions both as a data-collection tool and as a reporting agent that, in the longer term, may in fact be able to speak with humans about the value of its findings and how to reconfigure its efforts to provide needed data more effectively during team meetings. What's more, interacting with the learning machine, posing questions about the data it is generating and challenging racist, sexist and other problematic assumptions or outcomes embedded in or emergent from its data would all be more possible under the framework we are developing here. Data, like the machines that produce it, is not neutral and self-interpreting but should instead be seen as a product of team deliberation.[11]

Disease detection is also advancing through chatbots that provide individuals with information about illness. Babylon Health has developed a chatbot that works from an encyclopaedia of different symptoms to provide potential patients with a recommendation to consult with a physician, go to an emergency room, or determine if their condition does not require specialized care. Hamish Fraser, Enrico Coiera, and David Wong (2018) have raised concerns about programs such as the Babylon chatbot: 'Symptom checkers have great potential to improve diagnosis, quality of care, and health system performance worldwide. However, systems that are poorly designed or lack rigorous clinical evaluation can put patients at risk and likely increase the load on health systems' (p 2264). Topol (2019) echoed these concerns. Given that medical errors rank as the third leading cause of death in the United States behind heart disease and cancer (Mahajan, 2022), it is very important not to treat AI as a panacea. Indeed, as we accelerate into a world of AI, the transition should include safeguards for ethical practices and patient privacy.

Currently, physicians use AI to read and interpret medical scans and, in many cases, this represents an improvement over human scanning capability. As Charlene Liew (2020) opined:

> Current artificial neural networks have accuracy rates that surpass those of radiologists in narrow-based tasks such as nodule detection, and those of pathologists in detecting lymph node metastasis from breast cancer, and that are likely close to parity with ophthalmologists for vision-threatening diabetic retinopathy, with an area under the curve of 0.958 (95% confidence interval 0.956–0.961). (p 447)

In short, AI can provide a valuable resource for disease detection and will continue to develop in the coming years. However, seeing disease detection as a purely data-based endeavour would leave out the issue of illness care, as noted previously. Humanizing disease-detection software would mean having it learn not only from repeated scanning of signs and symptoms but also from the narrative and experiential reporting of patients themselves. A more humanized trajectory for AI in this regard would be to consider how computer programs (and later, robotic or android practitioners) might be trained to listen to human beings and how humans might involve themselves in the communicative and deliberative process of teaching computers to recognize not only disease but the characteristics of illness as a social, cultural, political and economic phenomenon that reflects the intersecting positionalities of race, gender, ability, sexuality and related elements of identity.

Distance medicine is an additional area where AI is teeming with possibility. Access to medical care in the developing world is fraught with complex problems. Shortages of doctors and health providers, the high costs of treatment and the lack of options in rural areas all contribute to zones of limited access. Chatbots provide a unique opportunity for revolutionizing distance medicine to address these challenges. Alan Greene, Claire C. Greene and Cheryl Greene (2019) have argued that chatbots can provide information to anyone with a mobile phone and 'can offer relevant high-quality information, reassurance, answers, and ways of thinking about the situation that might be more useful' (p 4). Drawing a parallel to self-driving cars, these authors noted that there will be a period of learning where human physicians or health providers must be present at each step because these decisions can mean life or death for the patients (see also Topol, 2019, pp 85–88). If the comparison is between access to a developing encyclopaedia of high-quality medical information or no information at all, the choice is clear. Imagine a world where everyone can instantly get information about how to deal with an ankle sprain, how to lower a fever, how to help someone who is having difficulty breathing. Knowing when a person absolutely needs emergency care and when treatment can mitigate symptoms is now an option. AI may be able to bring medical knowledge to those in peril from illness around the world, especially if it is adapted to work with rather than revise or replace extant local and indigenous ways of providing for health and wellness.

Moreover, with hospitals facing COVID-19 related crises including overcrowded emergency rooms and intensive care units, using AI to help patients determine what sort of treatment is appropriate and whether they absolutely need immediate emergency attention, a quick visit to a local clinic or can safely manage their situation from home will translate into more attention and care for the most serious cases. Chatbots being developed along these lines are also an opportunity to cultivate deliberative engagement between patients and AI, thus providing another avenue for (re)humanizing healthcare. In this case, patients can engage with AI as it learns how to best provide the most apposite and actionable information. Furthermore, AI can help patients to avoid the sometimes dehumanizing contexts of clinical care, especially when their conditions do not require a visit to a provider.

An additional area of interest is the use of AI in the field of pharmacy. Investment in AI by pharmaceutical companies is burgeoning. As David Freedman (2019) observed, 'AI-based drug-discovery start-ups raised more than $1 billion in funding in 2018, and as of last September, they were on track to raise $1.5 billion in 2019. Every one of the major pharmaceutical companies has announced a partnership with at least one such firm' (para 8). Freedman (2019) also noted that AI can assist with drug testing in a variety of ways, including 'by identifying more promising drug candidates; by raising the "hit rate," or the percentage of candidates that make it through clinical trials and gain regulatory approval; and by speeding up the overall process' (para 7). In addition to supporting research and development, chatbots can also deliver useful information on drug interactions to pharmacists (Greene et al, 2019).

In our view, AI can and should be configured as a deliberative partner with drug developers, pharmacists and patients; a conversation partner who can help them to see new opportunities to improve therapeutic agents and address the specific needs of patients. AI could collect data about patient experiences with drugs through pharmacists or the patients themselves, thus cultivating a deliberative network connecting the real-world use of drugs back to the pharmacists who deliver them and the scientists who develop them. This model of the healthcare team goes well beyond Gawande's (2011) initial vision but contains within it a powerful admixture of data made actionable through interactivity. Such interactivity would enhance the humanity of patients actually experiencing the effects of drugs and their pharmacists as legitimate practitioners and even contributors to research. It would also augment the humanity of AI as an interactive conduit that uses data to link patients to pharmacists, clinicians and researchers looking for ways to enhance the availability, acceptability and understanding of various drug regimens.

Crucially, the rhetorical-deliberative framework iteratively expounded across the various areas of AI development in the previous paragraphs may contribute to improved health communication and outcomes. A common refrain in health-communication studies is that good interactions between care providers and patients increase the likelihood of positive health outcomes. In their extensive account of health communication, Kevin B. Wright, Lisa Sparks, and Dan O'Hair (2008) detailed three health outcomes associated with successful communication between healthcare providers and patients: satisfaction with healthcare, adherence to treatment and improved physical and psychological health outcomes. High levels of patient approval, an increased likelihood that patients will carefully complete their treatment and better physical and mental health outcomes together provide a strong basis for understanding why health communication is a growing field. As digital beings become able to provide sophisticated diagnostics that significantly surpass the capabilities of physicians, the role of care will undoubtedly transition to one where the physician serves as a mediator between digital beings, patients and care teams. This will inaugurate a unique opportunity for improved health outcomes.

It is essential during this transition that physicians deepen their knowledge and ability to serve as leaders during team deliberations about care and expand their skill sets as communicators. The renowned communication scholars, Athena du Pré and Barbara Overton (2021), approached the study of health communication from an interpersonal or relational perspective that might enhance efforts to bring providers, patients and technology into productive dialogue with each other under the conditions we have been describing in this section. Du Pré and Overton (2021) identified five different kinds of relationships between providers and patients: 'mechanics'/'machines'; 'parents'/'children'; 'spiritualists'/'believers'; 'providers'/'consumers'; and 'partners' (pp 146–151). When care providers offer a diagnosis and treatment without direct input from the patient, the relationship is similar to a mechanic working on a vehicle. If the relationship between a care provider and a patient is asymmetrical, with the care provider guiding the treatment based on their knowledge and experience, this parallels the relationship between a parent and a child or even a believer and religious leader. When the patient takes on the role of customer, the model is based on the traditional conception of a consumer. Finally, if the patient and the care provider each contribute and share knowledge, the relationship is a partnership. While patient–provider relationships should always be taken as co-constitutive, du Pré and Overton (2021) introduced the idea that different kinds of communication create and sustain different relationships, some more deliberative, equitable and ultimately humanizing than others. Obviously, the idea of a partnership is much more compelling within our framework than any of the others, especially the thematically relevant notion of the mechanic and machine. But, the question arises, what should the contours of this partnership be and who/whom/what shall be included within it?

There are interesting synergies between du Pré and Overton's (2021) approach to thinking about patient–provider relationships and Gawande's (2011) notion of the healthcare 'pit-crew'. In this section, we explored a rhetorical-deliberative framework that includes providers, patients and AI on the healthcare team. Ample evidence demonstrates that AI is a useful repository of knowledge and provides a tool for quickly accessing information; however, it is crucial to acknowledge that digital beings can also assist with deliberations about care. Stanford's Center for Deliberative Democracy is at the time of writing using a chatbot to moderate deliberations by encouraging participation, transcribing dialogue in the moment and managing the agenda (Miller, 2020). In many ways, this serves a similar function to the 'checklists' for effective and ethical care advocated by Gawande (2009) in *The Checklist Manifesto* as a solution to the problem of complexity (see our introduction). As medicine becomes increasingly complex, the need for a systemic approach is essential to high-quality care, especially care of the sort that takes into account the whole lifeworld and experiential contexts of patients (Tronto, 2001; Gilligan, 2016). Team-based approaches that include AI participants can enhance the process of checking items off lists and addressing unexpected occurrences along the way. Of course, one of the major challenges for interdisciplinary health teams is effective group communication. As Wright, Sparks and O'Hair (2008) pointed out, 'Team members often require training in group communication and decision-making skills, especially since these skills are not always taught within the curriculum of some healthcare disciplines' (p 264). Chatbots such as the one utilized at Stanford's Center for Deliberative Democracy can assist with these interactions. But they can and should also be seen as actual participants in such deliberations, thus offering them opportunities to be humanized and, as such, to participate in the (re) humanization of the medical ecology.

In sum, and in all of the ways noted, AI can be seen as primarily a device for collecting and managing data. But, if we choose to configure AI as a deliberative participant on our healthcare teams, we may avoid two potential pitfalls of our technological era: (1) the view that decision making, whether medical or otherwise, is primarily a data-based affair with little room for argument between persons; and (2) the view that technology is somehow good because it lacks human characteristics and is, thus, less likely to make mistakes. Instead, we can and should see our technology as potentially ever more human and invested in the human elements of our lives, not just the pure data that might be collected from our blood, our vital signs or our standard symptoms. In short, if AI is to be a tool for (re)humanizing the medical ecology, it must be more than a machine. It must be seen as a being in the process of becoming, a potentially ever more human participant on our healthcare teams, capable of error but able to contribute in ways that are meaningfully complementary to other participants.

Conclusion

In his account of AI, Topol (2019) ultimately argued that doctors cannot and should not be *replaced* by AI because only they can truly provide the human dynamics of healthcare. Why? Because they are, in fact, human:

> The fundamentals – empathy, presence, listening, communication, the laying of hands, and the physical exam – are the building blocks for a cherished relationship between patient and doctor. These features are the seeds for trust, for providing comfort and promoting a sense of healing. They are the building blocks that enable genuine caring for the patient and a doctor's professional fulfillment that comes from improving a person's life. All these humanistic interactions are difficult to quantify or digitize, which further highlights why doctors are irreplaceable by machines. (Topol, 2019, p 302)

Central to his argument is that doctors are 'irreplaceable' for certain kinds of activities. AI can only replace doctors in the areas of diagnostics and some treatment decisions. This leaves open the possibility that, as AI advances, future doctors will not be prized so much for their medical knowledge as for their ability to be human beings with skills in ethical and communicative delivery of healthcare at those moments when we are truly in need of human contact.

From our perspective, Topol's argument opens the door for rhetorical scholars to make meaningful contributions to healthcare delivery. Indeed, we believe the discipline of communication can help to fill the gaps in medical education diagnosed in *Educating Physicians: A Call for Reform of*

Medical School and Residency (Cooke et al, 2010), a project supported by the Carnegie Foundation for the Advancement of Teaching. In that text, the authors argued that the older and very influential Flexner Report, also supported by this foundation (Flexner, 1910), had focused on standardizing scientific learning which, though important, is no longer the central challenge facing medical education. Instead, newer instructional methods and content focused on communication and ethics as central aspects of clinical care are beginning to take centre stage. These insights are now even more salient as the scientific skills originally prized at the beginning of the 20th century have increasingly become the domain of machines, leaving the affective, ethical, emotional and communicative tasks as the central activities of human practitioners. Medical schools and the host of programmes that serve the many allied health professions may shortly require fewer scientists and far more communication and ethics specialists as providers are prepared to enter the AI-enabled 21st-century clinic. For this reason, those who study human communication should be prepared to articulate how and why their disciplinary concerns are now central to the remaking of health praxis for providers and, we would add, for society writ large as the tools of care leak into the health apps and digital-medical ecologies of patients well beyond the clinical space (Lupton, 2016; Friz and Overholt, 2020).

Our argument, however, goes one step beyond Topol by seeking to understand not only how AI may make human beings more human but also how AI itself might be humanized. This line of thinking goes at least as far back as Alan Turing and his famous test. The 'Turing Test' (Richardson, 2015; Oppy and Dowe, 2021) was designed as a thought experiment and practical approach to thinking through the developmental pathways of AI. With this test, which focused on whether human beings could tell the difference between an AI and a human being, Turing forwarded the idea that at some point our machines would no longer appear to us as machines. We offer this not as a nightmarish, dystopian possibility but, rather, as the natural trajectory for a humanizing approach to technology. Instead of framing our technological ecology as dehumanizing, we have instead argued that a (re)framing of it as one more step towards a synergy between machine and human is the more helpful perspective. We have, thus, added the extra 'D' of *deliberation* to Topol's (2019, pp 15–16) transformational terms, hoping that in doing so we have carved out space for a different understanding of how we can and should interact with our electronic partners. AI should not merely be about information processing and delivery or correcting the error-prone human being. It should be a fully fledged discussant on Gawande's (2011) healthcare 'pit crew'. Human healthcare providers should avoid the 'cowboy' trope not only in terms of medical diagnosis and prognosis but also in the arena of patient-centred illness care. Every participant on the team should be part of this caring, up to and including AI.

And we mean this not only in the sense that AI might be able to contribute evidence-based practice to the caring endeavour (just as it may to the diagnostic and prognostic endeavours as noted earlier). We mean this in the sense that in humanizing AI and making it a deliberative *being* on the healthcare team, we may also open space to inform its more dramatic *becoming* as a rich partner in all of the activities of healthcare. Breaking down these barriers does not, in our view, risk the replacement of providers by machines. Instead, it allows machines to take their place with and next to providers in a medical ecology that is under-staffed, overworked and in need of new personnel with a variety of skills. We can see the value of this most clearly by imagining how AI could have been marshalled, if sufficiently advanced at the time, to provide care during this pandemic and during future disease outbreaks. Electronic caregivers would have the major benefit of being immune to emerging infectious diseases. They could, thus, be safely in the room with patients, helping them not only with the best treatments but also with the provision of companionship in a highly isolating context. This would protect human providers from infection and reduce the risk that they might bring infection into the room with them or pass it along to the next patient. In short, AI could be a deliberative and caregiving partner in future pandemics, one that could help individual patients avoid the isolation, disconnection and fear that were experienced so significantly during the COVID-19 pandemic.

The previous analysis may raise the question: what limitations, if any, should be placed on our reliance on AI? Should we draw lines between machines and humans and their role on the 'pit crew' (Gawande, 2011)? The lines between the epistemic and technical aspects of care and the affective, emotional, ethical and communication tasks of medicine are not, ultimately, as clear as Topol and others seem to suggest. The idea that caring is not potentially itself an evidence-based endeavour is belied by the increasing interest in patient-centred approaches to care in the status quo. What's more, the idea that machines are not beings and will never have the capacity to engage in human care work is a dehumanizing and anti-caring approach to the medical ecology that has grown up around us. While there are limitations to the power of words and ideas to remake the world around us, we choose to see AI as embedded in a process of becoming more human. We think this view is more productive than its opposite. Relying on human beings to be human and machines to be machines maintains a problematic and imbalanced conception of medicine. Collapsing the boundaries and seeing how far we may go in terms of realizing our version of a virtual doctor or a physically present practitioner, our own Data, will keep us focused on the endeavour to build worlds that are responsive to our needs and interests. As technology becomes sufficiently advanced, we may find that our creation comes home to roost not as an anti-human, data-crunching monster but as a being much more recognizable to us and, thus, more welcome at our moments of greatest need.

Notes

[1] We are using the term 'ecology' here to refer to the overall context of medicine – its institutions (including clinical sites and organizational networks reaching out into society), technological apparatus (including devices and virtual worlds), communicative styles and settings, practitioners and patients – which exist in productive tension with one another (for expressions of this view, see Peabody, 1927; Morris, 2000). That is, these elements intermingle, influence, intervene into and drive the evolution of each other both in terms of their make-up and, when it comes to the (sometimes, though not always as we shall see, human) beings involved, their perspectives. This is a slightly different use of the term than, for instance, has become important in medical anthropology and public health, where it has been used it to describe the interdependence of culture, disease and other factors in the overall cultivation of health and well-being (see, for example, McLeroy et al, 1988; McElroy and Townsend, 2015).

[2] Some of the biographical details noted here can be found in Gawande (2015).

[3] One of the authors, John Rief, would like to acknowledge and thank Kati Sudnick, PhD of the University of North Carolina Wilmington for her collaborative work with him while completing her degrees at Duquesne University in Pittsburgh, PA, to chart the development of patient-centredness as a concept and practice in contemporary healthcare. This work informs elements of the analysis in this chapter.

[4] For example, see Frank M. Snowden's (2020) discussion of the history of epidemic disease. From his perspective, epidemics cannot be understood through the narrow lens of pathogens. Instead, pathogens interact with bodies, environments and a variety of social, cultural, political and economic coordinates that lead to differential outcomes across time and around the globe.

[5] We are aware of the philosopher Annmarie Mol's (2002) efforts to trouble the hard and fast distinctions between disease (as scientific, biological, objective) and illness (as subjective experience) that can emerge from ideas like those shared in this section of the chapter. We agree with Mol's point. While we find it useful to use the term 'illness' to mark a patient's perspective on their disease, it is also important to note that disease is not a purely biological phenomenon, nor is it ontologically separate from illness. Indeed, this is precisely the point that Sacks (1983) made in his discussion of Pasteur, as previously noted. Disease is a product of its context. Following Mol, just as with illness, disease is culturally, socially, institutionally and technologically situated. And, disease and illness are bound together in each moment of the patient and practitioner encounter (on this, see Mol, 2002, pp 12–13).

[6] See especially Topol's (2019) discussion of nursing and at-home care in Chapter 7 of his book.

[7] Here, we are indebted to the work of Donna J. Haraway (2016), Distinguished Professor Emerita at the University of California, Santa Cruz, who has artfully deconstructed the boundaries between humans, animals and machines.

[8] The anthropologist Kathleen Richardson has investigated a variety of models for the realization of more or less *human* versions of AI. In *An Anthropology of Robots and AI: Annihilation Anxiety and Machines*, Richardson (2015) documents the many anxieties circulating around the construction of artificial beings that emerge both from popular culture and from the labs where they are being developed. Indeed, these domains are mutually informing in Richardson's account. Most importantly, Richardson analyses a widely shared concern about the potential for artificial beings to 'annihilate' humans

or, in the very least, render them indistinct from their creations. Richardson ultimately argues that machines cannot replace humans in the intimate relationships that are central to human interactivity and community. We tend more towards the boundary-crossing ideas of Haraway (2016), but take Richardson's point seriously. As we develop AI for health praxis, we should respect the limitations of technological change and the potential threats posed by the incursions of technology into human interactivity that can come from too eagerly and unreflectively crossing boundaries.

9   As with the other technological advances noted in this chapter, utilizing digital second opinions of this sort would require a device, access to high-quality internet and an understanding of how to use these tools in securing actionable health information. Thus, in addition to accessible design, numerous other concerns might be raised, including the ways in which class, gender, race and other modalities of difference and associated experiences of oppression and exclusion tend to exacerbate the digital divide. In our view, as expressed here and elsewhere, more work will need to be done to address these concerns in the ever more technologically enhanced 21st-century medical ecology. Indeed, such work is essential to avoid the promulgation of inequitable and outright oppressive ideologies under the mask of *neutral* technological tools (see, for example, O'Neil, 2016; Noble, 2018; Benjamin, 2019).

10 For similar arguments regarding the deliberative and contextual foundations of scientific knowledge made in the area of science and technology studies, see Helen Longino (1990). For a perspective focused on making space for traditionally marginalized ways of knowing in the (re)construction of scientific knowledge, see Sandra Harding (1986; 1991). We stand in agreement with Harding that science should involve multiple epistemologies that go beyond Gawande's concern with the inclusion of different kinds of practitioners (for example, surgeons, oncologists and so on), to include and value perspectives grounded in and responsive to sex, gender, sexuality, race, ethnicity and the like.

11 An additional challenge that emerges from the use of AI to collect data, detect disease and design effective public health interventions is the loss of privacy. In response to this problem, Topol (2019) noted the need for the development of methods and tools to protect patient privacy in the increasingly expansive and intrusive domain of Big Data (pp 101–104). While we do not have time to fully address these concerns here, it is crucial that ongoing deliberations about privacy protections inform the development of AI at every stage of the process.

References

Ahmed, S. (2002) 'This other and other others', *Economy and Society,* 31(4): 558–572.

Alsuliman, T., Humaidan, D. and Sliman, L. (2020) 'Machine learning and artificial intelligence in the service of medicine: necessity or potentiality?' *Current Research in Translational Medicine,* 68(4): 245–251.

Aristotle (1926) *Art of Rhetoric*, trans. J.H. Freese, Cambridge, MA: Harvard University Press.

Benjamin, R. (2019) *Race after Technology: Abolitionist Tools for the New Jim Code*, Cambridge: Polity Press.

Burke, K. (1954) *Permanence and Change: An Anatomy of Purpose* (3rd edn), Berkeley, CA: University of California Press.

Burke, K. (1966) 'Terministic screens', in *Language as Symbolic Action: Essays on Life, Literature, and Method*, Berkeley, CA: University of California Press, pp 44–62.

Chakravarty, K., Antontsev, V., Bundey, Y. and Varshney, J. (2021) 'Driving success in personalized medicine through AI-enabled computational modeling', *Drug Discovery Today*, 26(6): 1459–1465.

Clark, A.E. and Shim, J. (2011) 'Medicalization and biomedicalization revisited: technoscience and transformations of health, illness, and American medicine', in B.A. Pescosolido, J.K. Martin, J.D. McLeod and A. Rogers (eds) *Handbook of the Sociology of Health, Illness, and Healing: A Blueprint for the 21st Century*, New York: Springer, pp 173–199.

Comstock, J. (2018) 'Babylon's AI passes mock-up of UK's GP exam, goes head-to-head with doctors', *MobiHealthNews*, 27 June, www.mobihealthn ews.com/content/babylons-ai-passes-mockup-uks-gp-exam-goes-head-head-doctors.

Cooke, M., Irby, D.M. and O'Brien, B.C. (2010) *Educating Physicians: A Call for Reform of Medical School and Residency*, The Carnegie Foundation for the Advancement of Teaching, San Francisco, CA: Jossey-Bass.

du Pré, A. and Overton, B.C. (2021) *Communicating about Health: Current Issues and Perspectives* (6th edn), New York: Oxford University Press.

Ehninger, D. (1970) 'Argument as method: its nature, its limitations, and its uses', *Speech Monographs*, 37(2): 101–110.

Epstein, R.M. and Street, R.L. (2011) 'The values and value of patient-centered care', *The Annals of Family Medicine*, 9(2): 100–103.

Farrell, T.B. (1976) 'Knowledge, consensus, and rhetorical theory', *Quarterly Journal of Speech*, 62(1): 1–14.

Flexner, A. (1910) *Medical Education in the United States and Canada: A Report to the Carnegie Foundation for the Advancement of Teaching*, New York: The Carnegie Foundation for the Advancement of Teaching, http://archive. carnegiefoundation.org/publications/pdfs/elibrary/Carnegie_Flexner_ Report.pdf.

Frank, A.W. (2013) *The Wounded Storyteller: Body, Illness, and Ethics* (2nd edn), Chicago: University of Chicago Press.

Frank, D.A. (2004) 'Argumentation studies in the wake of *The New Rhetoric*', *Argumentation and Advocacy*, 40(4): 267–283.

Fraser, H., Coiera, E. and Wong, D. (2018) 'Safety of patient-facing digital symptom checkers', *The Lancet, [online]*, 392(10161): 2263–2264, https:// doi.org/10.1016/S0140-6736(18)32819-8.

Freedman, D.H. (2019) 'Hunting for new drugs with AI', *Nature', [online]*, 576: S49–S53, https://www.nature.com/articles/d41586-019-03846-0.

Friz, A. and Overholt, S. (2020) ' "Did you have sex today?": discourses of pregnancy and Big Data in fertility-tracking apps', in L. Melonçon, S.S. Graham, J. Johnson, J.A. Lynch and C. Ryan (eds) *Rhetoric of Health and Medicine As/Is: Theories and Approaches for the Field*, Columbus, OH: Ohio State University Press, pp 101–122.

Gawande, A. (2002) *Complications: A Surgeon's Notes on an Imperfect Science*, New York: Picador.

Gawande, A. (2009) *The Checklist Manifesto: How to Get Things Right*, New York: Metropolitan Books.

Gawande, A. (2011) 'Cowboys and pit crews', *The New Yorker*, 26 May, www.newyorker.com/news/news-desk/cowboys-and-pit-crews.

Gawande, A. (2014) *Being Mortal: Medicine and What Matters in the End*, New York: Metropolitan Books.

Gawande, A. (2015) 'Leading as a surgeon by day and writer by night', *Voices in Leadership*, Harvard, T.H. Chan School of Public Health [video interview], 21 April, www.hsph.harvard.edu/voices/events/gawande/.

Gawande, A. (2018) 'Why doctors hate their computers', *The New Yorker*, 5 November, www.newyorker.com/magazine/2018/11/12/why-doctors-hate-their-computers.

Gibbs, S. (2017) 'AlphaZero AI beats champion chess program after teaching itself in four hours', *The Guardian*, 7 December, www.theguardian.com/technology/2017/dec/07/alphazero-google-deepmind-ai-beats-champion-program-teaching-itself-to-play-four-hours.

Gilligan, C. (2016) *In a Different Voice: Psychological Theory and Women's Development*, Cambridge, MA: Harvard University Press.

Glouberman, S. and Zimmerman, B. (2002) 'Complicated and complex systems: what would successful reform of Medicare look like?' Commission on the Future of Health Care in Canada, www.alnap.org/system/files/content/resource/files/main/complicatedandcomplexsystems-zimmermanreport-medicare-reform.pdf.

Greene, A., Greene, C.C. and Greene, C. (2019) 'Artificial intelligence, chatbots, and the future of medicine', *The Lancet Oncology*, 20(4): 481–482.

Groopman, J. (2007) *How Doctors Think*, Boston, MA: Houghton Mifflin Company.

Gross, A.G. (1996) *The Rhetoric of Science*, Cambridge, MA: Harvard University Press.

Hamburg, M.A. and Collins, F.S. (2010) 'The path to personalized medicine', *New England Journal of Medicine*, 363(4): 301–304.

Haraway, D.J. (2016) 'A cyborg manifesto: science, technology, and socialist-feminism in the late twentieth century', in *Manifestly Haraway: The Cyborg Manifesto, The Companion Species Manifesto, Companions in Conversation (with Cary Wolfe)*, Minneapolis, MN: University of Minnesota Press.

Harding, S. (1986) *The Science Question in Feminism*, Ithaca, NY: Cornell University Press.

Harding, S. (1991) *Whose Science? Whose Knowledge? Thinking from Women's Lives*, Ithaca, NY: Cornell University Press.

Hiland, E.B. (2021) *Therapy Tech: The Digital Transformation of Mental Healthcare*, Minneapolis, MN: University of Minnesota Press.

Institute of Medicine (2001) *Crossing the Quality Chasm: A New Health System for the 21st Century*, Washington, DC: National Academy Press.

Kitcher, P. (1995) 'The cognitive functions of scientific rhetoric', in H. Krips, J.E. McGuire and T. Melia (eds) *Science, Reason, and Rhetoric*, Pittsburgh, PA: University of Pittsburgh Press, pp 47–66.

Kleinman, A. (1988) *The Illness Narratives: Suffering, Healing, and the Human Condition*, New York: Basic Books.

Kuehl, R., Drury, S. and Anderson, J. (2020) 'Rhetoric *as* rhetorical health citizenship: rhetorical agency, public deliberation, and health citizenship *as* rhetorical forms', in L. Melonçon, S.S. Graham, J. Johnson, J.A. Lynch and C. Ryan (eds) *Rhetoric of Health and Medicine As/Is: Theories and Approaches for the Field*, Columbus, OH: Ohio State University Press, pp 161–181.

Laine, C. and Davidoff, F. (1996) 'Patient-centered medicine: a professional evolution', *Journal of the American Medical Association*, 275(2): 152–156.

Latour, B. (1987) *Science in Action: How to Follow Scientists and Engineers through Society*, Cambridge, MA: Harvard University Press.

Liew, C. (2020) 'Medicine and artificial intelligence: a strategy for employing Porter's classic framework', *Singapore Medical Journal,* 61(8): 447.

Longino, H.E. (1990) *Science as Social Knowledge: Values and Objectivity in Scientific Inquiry*, Princeton, NJ: Princeton University Press.

Lupton, D. (2016) *The Quantified Self: A Sociology of Self-Tracking*, Cambridge: Polity Press.

Mahajan, P. (2022) *Artificial Intelligence in Healthcare* (2nd general edn), Albuquerque, NM: MedMantra, LLC.

McElroy, A. and Townsend, P.K. (2015) *Medical Anthropology in Ecological Perspective* (6th edn), New York: Routledge.

McGrail, K.M., Ahuja, M.A. and Leaver, C.A. (2017) 'Virtual visits and patient-centered care: results of a patient survey and observational study', *Journal of Medical Internet Research*, 19(5), www.ncbi.nlm.nih.gov/pmc/articles/PMC5479398/.

McLeroy, K.R., Bibeau, D., Steckler, A. and Glanz, K. (1988) 'An ecological perspective on health promotion programs', *Health Education Quarterly*, 15(4): 351–377.

Miller, K. (2020) 'A moderator chatbot for civic discourse', https://hai.stanford.edu/news/moderator-chatbot-civic-discourse.

Mitchell, G.R. (2006) 'Team B intelligence coups', *Quarterly Journal of Speech*, 92(2): 144–173.

Mitchell, G.R. (2010) 'Switch-side debating meets demand-driven rhetoric of science', *Rhetoric & Public Affairs*, 13(1): 95–120.

Mitchell, G.R. (2021) 'Debating with robots: IBM Project Debater and the advent of augmentive automated argumentation', in T. Suzuki, T. Tateyama, T. Kato and J. Eckstein (eds) *Proceedings of the Tokyo Conference on Argumentation, Volume 6*, pp 91–102, https://japan-debate-association.org/en/proceedings-tca-vol-6.

Mitchell, G.R. and McTigue, K.M. (2012) 'Translation through argumentation in medical research and physician-citizenship', *Journal of Medical Humanities*, 33(2): 83–107.

Mol, A. (2002) *The Body Multiple: Ontology in Medical Practice*, Durham, NC: Duke University Press.

Montgomery, K. (2006) *How Doctors Think: Clinical Judgment and the Practice of Medicine*, New York: Oxford University Press.

Montgomery, S.L. (1995) 'Illness and image: an essay on the contents of biomedical discourse', in *The Scientific Voice*, New York: The Guilford Press, pp 134–195.

Morris, D.B. (1993) *The Culture of Pain*, Berkeley, CA: University of California Press.

Morris, D.B. (2000) *Illness and Culture in the Postmodern Age*, Berkeley, CA: University of California Press.

Morse, J.M. and Johnson, J.L. (eds) (1991) *The Illness Experience: Dimensions of Suffering*, Newbury Park, CA: Sage.

Noble, S.U. (2018) *Algorithms of Oppression: How Search Engines Reinforce Racism*, New York: New York University Press.

O'Neil, C. (2016) *Weapons of Math Destruction: How Big Data Increases Inequality and Threatens Democracy*, New York: Crown.

Oppy, G. and Dowe, D. (2021) 'The Turing Test', *Stanford Encyclopedia of Philosophy*, https://plato.stanford.edu/entries/turing-test/.

Peabody, F. (1927) 'The care of the patient', *Journal of the American Medical Association*, 88(12): 877–882.

Pellegrino, E.D. (2001) 'The internal morality of clinical medicine: a paradigm for the ethics of the helping and healing professions', *The Journal of Medicine and Philosophy*, 26(6): 559–579.

Pierret, J. (2003) 'The illness experience: state of knowledge and perspectives for research', *Sociology of Health & Illness*, 25(Silver Anniversary Issue): 4–22.

Richardson, K. (2015) *An Anthropology of Robots and AI: Annihilation Anxiety and Machines*, New York: Routledge.

Rief, J.J., Hamm, M.E., Zickmund, S.L., Nikolajski, C., Lesky, D., Hess, R. et al (2017) 'Using health information technology to foster engagement: patients' experiences with an active patient health record', *Health Communication*, 32: 310–319.

Sacks, O. (1983) *Awakenings* (revised edn), New York: E.P. Dutton.

Slonim, N. et al (2021) 'An autonomous debating system', *Nature*, 591(March): 379–384.

Snowden, F.M. (2020) *Epidemics and Society: From the Black Death to the Present*, New Haven, CT: Yale University Press.

Sprague, R.K. (2001) *The Older Sophists*, Indianapolis, IN: Hackett Publishing Company.

Stanford Health Care (2021) 'Stanford Medicine Online Second Opinion', https://stanfordhealthcare.org/campaigns/stanford-medicine-online-second-opinion.html.

Star Trek: The Next Generation (1987–94) Paramount Domestic Television.

Topol, E. (2013) *The Creative Destruction of Medicine: How the Digital Revolution Will Create Better Healthcare*, New York: Basic Books.

Topol, E. (2016) *The Patient Will See You Now: The Future of Medicine is in Your Hands*, New York: Basic Books.

Topol, E. (2019) *Deep Medicine: How Artificial Intelligence Can Make Healthcare Human Again*, New York: Basic Books.

Tronto, J.C. (2001) 'An ethic of care', in M.B. Holstein and P.B. Mitzen (eds) *Ethics in Community-Based Elder Care*, New York: Springer, pp 60–68.

Verghese, A. (2019) 'Foreword', in E. Topol (author) *Deep Medicine: How Artificial Intelligence Can Make Healthcare Human Again*, New York: Basic Books, pp ix–xiii.

Wieczorek, S.M. (2020) *eMessaging and the Physician/Patient Dynamic: Practices in Transition*, Lanham, MD: Lexington Books.

Winderman, E. and Landau, J. (2020) 'From HeLa cells to Henrietta Lacks: rehumanization and pathos as interventions for the rhetoric of health and medicine', in L. Melonçon, S.S. Graham, J. Johnson, J.A. Lynch and C. Ryan (eds) *Rhetoric of Health and Medicine As/Is: Theories and Approaches for the Field*, Columbus, OH: Ohio State University Press, pp 52–73.

WorldCare (2017) *The Powerful Impact of Medical Second Opinions*, www.worldcare.com/2017/07/31/powerful-impact-medical-second-opinions/.

Wright, K.B., Sparks, L. and O'Hair, H.D. (2008) *Health Communication in the 21st Century,* Malden, MA: Blackwell Publishing.

Zeng, D., Cao, Z. and Neill, D.B. (2021) 'Artificial intelligence-enabled public health surveillance- from local detection to global monitoring and control', in L. Xing, M.L. Giger and J.K. Min (eds) *Artificial Intelligence in Medicine: Technical Basis and Clinical Applications*, London: Academic Press, pp 437–453.

6

Digital Health Technological Advancements and Gender Dynamics in STS

Anamika Gulati

Introduction

I contend that the future of healthcare lies in the advancement of digital health technologies. Digital technologies such artificial intelligence, robotics, nanotechnology, virtual reality and augmented reality systems are changing how we practise and experience healthcare delivery and will challenge the traditional system of medicine.

A new technology emerges every day and is having substantial impact on the workforce. Due to the contrast between the scarcity era in which we have lived and the coming era of abundance through new technology, it is important to pay attention to social contracts. Adapting healthcare systems and policies to the development, assessment and uptake of health technologies is discussed in this chapter. Furthermore, this chapter highlights new and emerging opportunities for policy makers, as well as current challenges. It has been observed that these technological advancements have not benefited all classes of society and thus there are economic and cultural inequalities (World Social Report 2020; UNCTAD 2021). In this chapter, the reasons for the gap between classes of Indian society will be discussed and evaluated, as well as gendered asymmetries in access to healthcare services and technology. My focus is to identify the gender inequalities, associated risks and the developmental interventions that can fill this wide gap. Although advancements have been made in STEM (science, technology, engineering, and mathematics) areas by women in the past decade that has shaped contemporary science and technology studies (STS), this arena has not been left untouched by the issue of gender bias.

Further, the chapter will engage with knowledge communication and try to assess how the scientific information route is being impacted upon by the gender issue. Transmission of healthcare information to requisite audiences occurs through mediated or unmediated modes of communication. Diverse methodologies employed for the internalization of healthcare information, including edutainment or civic journalism, for example (Dunwoody, 2014; Parvanta 2020). This scientific information has to penetrate to every level of society, including socially and economically excluded groups who have worse health outcomes (Tarango and Machin-Mastromatteo, 2017). Several areas of health information are relevant to health literacy, such as improving clinical relationships between healthcare professionals and patients, enhancing compliance with clinical recommendations, developing public health campaigns, communication about individual and population health risks through mass media, the image of health in popular culture, educating the public about how to utilize the healthcare system and the development of telehealth applications. However, as in the STEM fields, a gender bias has also been reported in science communication. This chapter focuses on the representation, communication and internalization of relevant health knowledge, associated gender biases and how scholars have grappled with these biases in various areas of science communication.

Digital science and gender bias

Penetration of medical technologies

The healthcare industry has contributed greatly to society's health, extending life expectancy and increasing quality of life. Healthcare is being reshaped by emerging technologies in numerous ways – from how consumers gain access to it, how and which providers deliver it and how it ultimately impacts on outcomes. Healthcare is now distinguished by a modernized infrastructure for handling data and transactions, more efficient healthcare supply chains, better and faster development of therapies and intelligent and personalized healthcare ecosystems.

New medical technologies such as remote sensors, robotics, genomics, stem cells and artificial intelligence (AI) are rapidly gaining traction. Technology in the field of digital health is on the verge of becoming normalized and is likely to change the way healthcare is delivered. It will put the age-old and deeply rooted traditional medical system to the test.

Scientific advances have led to significant advances in understanding disease and developing better treatments. Several factors have contributed to the transformation of healthcare delivery by technology, including increasing the number of evaluable conditions and patient types, substituting existing treatments and methodologies or targeting them more specifically, intensifying treatments for specific conditions and changing delivery systems

(Taylor, 2015; Litwin, 2020). Medical and technological knowledges have been amalgamated through the development of nanotechnology and other digital tools. We have mobile applications for virtually everything, such that these technologies permeate every aspect of our lives. Digitalization has made it possible for us to access information that is needed to maintain our bodies properly and stay fit longer, whether through gaming, economics or fitness. Online doctor consultations (mHealth) have become more popular in the last 10 years (Devraj, 2019; Haleem, 2021). mHealth provides fast, accurate and comprehensive suggestions and methods on how to address specific health concerns. Having appointments online eliminates the inconvenience of having to visit a healthcare provider's office. Through online consultations, medical professionals and patients are connecting more quickly, since physical barriers are removed. An online consultation provides a more diverse space for knowledge and the practice of medicine that isn't limited by geographical location (Ganasegeran, 2019; Almathami 2020). However, the advent of mHealth has been associated with some limitations such as poor body language and communication (for example, lack of eye contact and physical contact) between the patient and the physician which may hinder quality online consultations. Additionally, online consultations aren't popular with many patients because they may be unfamiliar with the technology, and apprehensive about technical skills and the stability of internet connections (Vaportzis et al, 2017; Knowles and Hanson, 2018). It is possible for patients with complex medical conditions to need online consultation systems that include vital sign monitoring sensors linked to the data centre of their healthcare provider, or they may be required to periodically submit their data. Using these complex systems may cause difficulties, and they are not easy to use. Due to this, patients are less likely to accept online consultation systems and find them less convenient, which increases their reluctance to use them (Kruse, 2019; Almathami 2020). The evidence for the efficacy of mHealth (for example, to improve health outcomes) continues to grow, but there are concerns about the technology that may impair its scalability, effectiveness and, ultimately, the benefits of mHealth to public health. Slovensky and Malvey (2014) reviewed a number of global health policy expectations in the mHealth arena. The authors predicted that there would be an increase in public–private partnerships affecting patients, governments, entrepreneurs and developers. Unfortunately, this prediction has not been borne out. The current state of health policy appears to lag behind the potential for increased use of mobile health technologies and often serves as a hindrance rather than a facilitator for optimizing them (Ku and Sim 2021).

In general, the development and application of medical technology has resulted in longer life expectancy and survival rates, primarily as a result of the resources dedicated to its development (Kruk, 2018; Wamble, 2019). There are many limitations to this type of research, including suggested relationships between technologies and health in relation to factors unrelated to medicine, and a lack of quality data that includes more than just mortality, such as quality of life and function. Healthcare as a field often mistakenly sees technological innovation as incompatible with the philosophy of holistic care. As a result, technology is perceived as artificial, manufactured and impersonal – as valuing human experience only in terms of developing better algorithms and treating a person's body without regard for their identity. In contrast, holistic care focuses on the individual's physical, emotional and spiritual needs and sees each person not as potentially diseased but as a unique individual with a singular background and culture. Escalating expenditures on technology-enabled therapies in the last 10 years may not be matched by corresponding health gains. Additionally, contemporary technological advances are increasingly becoming a part of daily life and becoming ever more entwined with previous inventions and innovations. Technological advances are transforming our lives such that they are becoming heavily dependent on machines and data. Although combining technology and human lifestyles can have a positive effect on health in some ways, it can also have negative impacts. In the recent past, we have observed huge variations in intensive care admissions which have little effect on clinical outcomes but result in significant cost inflation. According to studies, end-of-life invasive medical treatments can be financially toxic, providing little assistance to patients and their families, in addition to causing suffering and distress (Cardona-Morrell, 2016; Carrera, 2018). Another example of this is the rise in antimicrobial resistance, caused in part by the unrestrained use of antibiotics. A greater effort could have been made to ensure that antibiotics are used appropriately and prudently on both the human and agricultural levels, so that the world would not now be facing the costs of high antimicrobial resistance. Also, these new technologies are blurring the line between medical devices and medicines, and they incorporate digital technology which requires new regulatory pathways. Moreover, the prices of many drugs on the market are astronomical, making treatment difficult to attain and putting the current health financing model at risk (Cardona-Morrell, 2016; Carrera, 2018).

Big Data in public health

Although technology has penetrated deeply into the healthcare system and Big Data and information technologies have been used for public health purposes such as research, health system improvements and surveillance of disease, there is still a lot more to be done in many countries to establish laws and policies that will enable the use of technological health and healthcare data in a secure way. In healthcare systems, technology is considered valuable only when the benefits outweigh the costs. We can only achieve this if we encourage access to safe, efficient, effective and clinically valuable technologies, and if we use them appropriately. Market entry regulations for new digital health technologies are essential to ensure consistent access to promising pharmaceuticals that address unmet medical needs without adversely impacting on patient safety (Mathews, 2019; Horgan, 2020). Products with incomplete pre-market evidence should be adequately disclosed to patients as quasi-experimental products. In addition, medical devices should be more tightly regulated to improve their performance and safety – especially those associated with high patient risks, which need a system to identify problems post-launch. This requires utilizing health data from across the country and regions as well as monitoring and sharing information about how medical devices perform in routine clinical settings. As mHealth products become more popular, a regulatory framework is essential to ensure their safety, manage privacy and security risks, encourage high-value innovation and prevent low-value, ineffective and unsafe products from swamping the market and displacing more beneficial products (Wallis, 2017; Jogova, 2019).

Digital health knowledge economy

Knowledge asymmetries between clinicians and patients in the health sector can be reduced by the digital health knowledge economy. Access to the digital health knowledge economy is restricted by information disparities between users and technology providers. In the future, we need to develop and apply technology intelligently, on the basis of evidence, with a focus on improving both individual and population health through sound policy frameworks (WHO, 2001; WHO, 2021). According to a report on Global Strategies on digital health by the WHO, 'Digital health should be an integral part of health priorities and benefit people in a way that is ethical, safe, secure, reliable, equitable and sustainable. It should be developed with principles of transparency, accessibility, scalability, replicability, interoperability, privacy, security and confidentiality' (WHO, 2021). As more sophisticated and more costly technologies become available, the sustainability of healthcare systems will become more difficult, necessitating a rationally deployed approach to treatment, interventions and healthcare system resources.

The future of healthcare will be determined by technological advancements more than by any other factor, and it will continue to evolve at a dramatic pace. It's possible and worthwhile to speculate about future trends in healthcare, but it's equally important to grasp the underlying drivers of progress to align with and work to achieve the best outcomes for society. The European Observatory on Health Systems and Policies by the WHO emphasizes that a regulator or policy maker should evaluate technology based on how it balances improvement in longevity with improvement in quality of life, stating:

> Health technologies pose similar challenges to health-care systems throughout the world. Thus, it is necessary to ensure that health technologies are evaluated properly and applied to health care efficaciously. In order to optimize care using the available resources, the most effective technologies should be promoted while taking consideration of organizational, societal and ethical issues. (WHO, 2008)

Moreover, health policies should be rotated so that they consider improving the health of the population within a budgetary and regulatory framework. Achieving this goal calls for policy makers to promote the development and adoption of technologies that improve the health of populations, with a focus on improving patient quality of life, ensuring they are accessible to all and promoting sustainable healthcare systems (Neumann and Weinstein. 1991; WHO, 2008; OECD, 2012).

Health technologies and gender dynamics

An additional issue that prevents health technologies from being used effectively is gender bias. Relationships between patients and healthcare providers affect the quality of care that patients receive and should be defined by respect, openness and balance in providers' decision-making roles. Segregation and inequality have long plagued the world of science and technology. Women are discriminated against in the nature and style, content and practices of medical care. According to Regitz-Zagrosek (2012), 'the prevention, management and therapeutic treatment of many common diseases does not reflect the most obvious and most important risk factors for the patient: sex and gender'. Healthcare technology has progressed by including AI technologies; however, sex and gender differences and biases are still observed as rightly highlighted by Cirillo et al (2020). Further, Ernst (2014) explored the role gender plays in scientific research and the development of new technologies, and suggested that researchers disenfranchised by gender hierarchies, racism, classism, homophobia and other ideological frames to categorize persons in structural hierarchies can be empowered through feminist analysis of science and technology. Techno-scientific advances are radically transforming the interactions between women and machines, argues Wajcman. But it's feminist politics rather than technology itself that makes a difference. In techno-feminism, visionary insights from cyberfeminism are combined with a materialist analysis of the sexual politics of technology. Wajcman explores how technology is gendered both in its design and use, utilizing new perspectives in postmodernism, feminist theory and STS.

Meanwhile, she shows how the technological culture of the world in which we live shapes our very subjectivity (Wajcman, 2004).

As an example of advancing healthcare technologies, health robotics are becoming increasingly impactful, in particular in neurology, rehabilitation and assistive approaches for improving the quality of life of patients and caregivers. The issue of how a robot may be adapted to a person's gender (for example, speech style) has not received a lot of research, but a few studies have looked at the effects of gender on human–robot interactions. For instance, in a recent study, sex differences in how children interact with robots were revealed, with implications for their use in paediatric hospitalization (Kory-Westlund, 2019; Logan, 2019). Education, employment and the health of women and men have all made significant progress in the past few decades, but health disparities continue to exist. There have always have been obstacles to access for women because of gender bias. In many cases, patients perceive the patient–provider relationship as discriminatory, marginalizing, abusive, as well as perpetuating societal stratification. Developing countries as well as developed ones are affected. Healthcare is often poor and inadequate for poor women and men, due to discrimination and class differences, which are often mediated by ethnicity, religion, language and so on (Govender and Penn-Kekana, 2007; Rivenbark and Ichou, 2020; Baciu 2017).

Health information is widely available online, but information overload makes it difficult for people to parse it effectively. By disintermediating data and information, users can become more informationally autonomous, but navigating the complex ecosystems into which they are embedded can still be very challenging (Laugksch, 1999; DeBoer, 2000; Yanyan, 2021). This is especially the case for communities where structural inequalities restrict opportunities to acquire the knowledge one needs to distinguish between credible and non-credible sources. South Asia has the highest gender gap, with 51% or 300 million fewer women using internet-connected mobile phones as compared to men by the end of 2020 in low- and middle-income countries (GSMA Report, 2020). According to a report published by UNESCO, skill deficiencies have contributed more to the widening of digital gender gaps than issues with access. This may be explained by a variety of factors, such as prohibitively high prices, a lack of educational opportunities and a lack of relevant and empowering content (UNESCO, 2017; OECD, 2018).

Researchers have found that women and men communicate and receive healthcare services differently, but no study has examined how power dynamics affect access to healthcare technologies between the patient and healthcare workers. The evidence-based care received by women with the same condition may differ from that received by men. Women are treated differently in several critical areas, such as cardiology and pain management, which can lead to poorer outcomes (Vlassoff, 2007; Samulowitz, 2018;

Woodward, 2019). Further, patients' communication with healthcare workers may also be fundamentally affected by gender dynamics. This may be in regard to a sexually transmitted infection (STI) or to something like pain, where studies have shown that various cultural norms apply to how much pain is acceptable and women are usually asked to minimize the expression of pain (Bendelow, 1993; Govender and Penn-Kekana, 2007). The world over, health reforms in the past few decades have largely focused on providing services in underserved areas and providing financial assistance to the poor. Gender equality is an enshrined fundamental right in the preamble of the Indian Constitution. India has ratified a number of international conventions and human rights forums (for example, Convention on the Elimination of All Forms of Discrimination against Women, Night Work [Women] Convention and so on) to guarantee women's equality and eliminate all forms of gender discrimination.

Digital health and user education

There is another group that is negatively affected by the advent of digital health technologies, and this is the digitally uneducated population, as they cannot verify the accuracy of information coming from online sources. It seems that a correlation exists between low levels of eHealth literacy and reliance on poor-quality sources of health information, so these groups are more likely to be targeted by misinformation (Eysenbach, 2008; Lupton, 2014). It is imperative that companies operating in this space take great care when collecting, processing and storing health data, even though these digital health technologies can provide easy access to healthcare services in terms of both treatment and preventative care. Sharing health data can lead to innovations that transform patient care, but only if it is handled responsibly. This is particularly the case since individuals' privacy can be seriously impacted by the collection and use of health data. Pritts (2008) discusses the high value placed on the on privacy, confidentiality and security of health information. Further, there is a school of thought that views privacy as an inherent human right with intrinsic value (Fried, 1968; Moore, 2005; NRC, 2007; Terry and Francis, 2007). Technology is transforming health systems into 'honey pots' (that is, storing *all* of an individual's personal information). Since health records are stored in one place, this poses serious issues of trust and security (IMC, 2009; Abouelmehdi, 2018). Additionally, such actions can undermine public trust in governments and social institutions (OECD, 2013). Thus, policy makers must manage the risks in a way that maintains public trust while also promoting economic growth in order to optimize the outcomes and benefits of treatments under assessment. Further, health information assets that are not developed, are unused or hard to use pose significant risks to individuals and societies. When assessing risks, it is essential to take both societal and individual perspectives into account (Guide to Privacy and Security of Electronic Health Information, 2015).

Data security

Healthcare data can be used in line with privacy preservation and utilization (OECD, 2013), using a variety of privacy-enhancing technologies. A number of different approaches to data anonymization have been developed, including de-identification and pseudonymization (OHRP, 2008; Clifton, 2013; PTAC, 2013; Cavoukian, 2014). In the process of de-identification, patients lose certain identifying information, such as their names, telephone numbers, addresses and critical dates. Meaningless codes can be substituted for key patient identifiers by pseudonymizing patient data, allowing re-identification for approved purposes. Before data are made available for analysis and research, it is critical that these practices be implemented, but in a way that accounts for any unintended consequences. Data de-identification techniques, however, usually do not completely eliminate the possibility of a dataset being worked or combined with other data to reveal the identity of the subject (OECD, 2015; OECD, 2020). This presents a particularly challenging situation, since removing all identification risks from Big Data is becoming increasingly difficult. Big Data can be viewed as both utopian and dystopian, like other sociotechnical phenomena (Boyd and Crawford, 2012; Shin and Choi, 2015). There are a number of fields where Big Data may help, such as cancer research, terrorism and climate change. Alternatively, it can also be used as a means of addressing societal issues. Meanwhile, Big Data is considered a troubling manifestation of Big Brother, allowing intrusions into personal privacy, reducing civil liberties, widening inequality and increasing corporate and state control. All sociotechnical shifts are characterized by hope and fear, which obscure subtler and more nuanced shifts that are occurring. Data collected by nations is likely to yield the highest economic benefits. India's 1.3 billion people, for example, create vast amounts of data every day, offering an opportunity to harness and manage the data behemoth for the country's benefit through Big Data analytics.

However, to be able to reap the benefits of safe data use, a data governance framework with mechanisms and best practices is needed at each stage of data development and use. Contract agreements require data receivers to follow security and disclosure practices, and independent bodies evaluate data-use proposals for public benefit and security adequacy. Compliance is monitored through regular security audits and follow-up actions (OECD, 2013; OCED, 2015). Under the provisions of the Information Technology Act 2000 and the Information Technology Rules 2011, certain information, including medical records, is protected from collection, disclosure and transfer under Indian law. Although many aspects of the law have been resolved, much of the legislation has not kept pace with technological advancements. Accordingly, the government has established the Digital Information Security in Healthcare Act (DISHA) and the Personal Data Protection Bill 2019 (PDP Bill). The PDP Bill covers the collection, disclosure, sharing and processing of personal data in India by the state, Indian companies, Indian citizens or any other entity incorporated or created in India.

Electronic medical records (EMRs) and electronic health records (EHRs) have become the preferred method of storing patient information in India. The Clinical Establishments (Registration and Regulation) Act 2010 requires an EMR to be maintained and provided to each patient by the clinical establishment. Moreover, the Ministry of Health and Family Welfare (MoHFW) was the first government agency to introduce EHR Standards, a standard-based system for creating and maintaining electronic health records by healthcare providers. These provisions were made public in December 2016 and stated that individuals must be informed about both immediate and future reasons for using their data, within the context of informed consent (Singleton, 2006; Ploug, 2020). However, the data contributed by an individual can sometimes be restricted or withdrawn from statistics and research. In some cases, this may be acceptable, but in others it may undermine the purpose which the data is being used and may result in biased results, with potential societal repercussions. It is extremely important to take care when withdrawing or restricting data to avoid compromising its integrity or the data's use for research. Consent mechanisms play an increasingly important role in data governance as technology and the law evolve (OECD, 2013a; 2013b; 2015).

Asymmetries in technological advancements

These technological advancements and research in various biomedical fields have not benefited all classes of society and thus there are economic and cultural inequalities that have resulted (World Social Report 2020; UNCTAD 2021). The following section discusses the gendered asymmetries in the role of women in technology development and growth of women in the STEM arena worldwide, with a special focus on Indian industries. Since the early 1970s, women and girls have made significant progress in education and workforce participation. However, men still outnumber them, especially in the upper echelons and certain science and engineering fields (Funk and Parker, 2018; Stewart-Williams and Halsey, 2021). The STEM fields have been criticized for the gender imbalance among students who graduate with these degrees and pursue careers in these fields. Women have a significantly lower likelihood than men of graduating with a STEM degree or working in STEM fields. The likelihood of women enrolling in a STEM programme immediately after high school graduation is 29.8% lower than for men. Focusing on bachelor's degree STEM programmes, the gender gap in enrolment falls to 19.9%; however, it almost doubles (to 36.4%) when high school graduates who have enrolled in a bachelor's degree STEM programme are considered (Chan, 2021; Vooren, 2022). The gender gap in STEM leads not only to unfair treatment of women in these fields, but also to a decline in the quality of work and innovation. The advancement of science requires diverse perspectives, and closing the gender gap and making the STEM fields more diverse will ensure that tomorrow's scientists will approach problems from a wide range of perspectives, as is argued by Haraway (1988). STS emerged during the second wave of women's liberation. By proposing feminist criteria for producing knowledge, design and technologies with a focus on social justice, STS feminists aim to reshape epistemologies, methodologies and political futures. This is illustrated by research in the natural sciences – specifically physics and biology (Haraway, 1988; Barad, 1995) – the use of standards (Star, 1991) and prenatal screening (Rapp, 2000). Following these inquiries into science and the use of technologies, feminist methodologies have been rethought (Harding, 1986; Haraway, 1988; Lykke, 1996; Barad 1995; 2007). The emergence of new reproductive technologies has prompted many other theoretical developments in gender studies and feminist STS, due to the connections between science, theory and politics (Adrian, 2014). Globalized health politics have infused these agendas into feminist activism on reproductive health matters (Murphy, 2012). Activist feminists began criticizing science and technology at the same time as women within the natural sciences began challenging science's premises (Fausto-Sterling, 1985, Trojer, 1985; Birke, 1986; Haraway, 1987; 1988; Keller, 1992; Oudshoorn, 1994; Barad, 1995). As part of this epistemological critique, the dichotomy between discourse and materiality has been challenged. The goal of gender equality in science, technology and innovation goes beyond just fairness; rather than just improving opportunities for women, gender equality aims to promote scientific and technological excellence and social justice (genSET, 2011). The STEM workforce will become more innovative, creative and competitive if there are more women, which is necessary in light of the challenges facing scientists and engineers today, such as global warming, the development of renewable energy sources and understanding the origins of the universe, among other matters.

In most cases, however, women are left out of the loop when it comes to the design of these products. Acoustic-recognition systems, for example, were originally calibrated for hearing the voices of men, and thus women's voices were not recognized. In addition, several other industries suffer from the same issue. One example is the first generation of automotive airbags, which were designed by a male-dominated engineering team, resulting in women and children being killed, whose deaths might otherwise have been prevented (Margolis and Fisher, 2002). The analysis revealed that development teams – most of whom were men – have always used male crash-test dummies! In the past, there were no female crash-test dummies at all, something that the design teams evidently didn't notice for quite some time. Other examples include the development of the artificial heart valve (Margolis and Fisher, 2002). Based on US Census Bureau data since 1990, the Pew Research Center found that although the number of STEM jobs has increased substantially, particularly in computer occupations, the proportion of women in STEM jobs has not crossed the 25% mark by 2016 (Funk and Parker, 2018). There is considerable variation in the percentage of women in the 74 STEM occupations that were studied – from only 7% of sales engineers and 8% of mechanical engineers to 96% of speech therapists and 95% of dental hygienists. Women make up an overwhelming majority of workers in health-related occupations, but only 14% of engineers, on average. Through a US Census Bureau data analysis, it was observed that a group of computer occupations that includes computer scientists, systems analysts, software developers, information systems managers and programmers – the STEM occupation cluster that has become increasingly popular since the 1990s – has actually seen a decrease in female gender representation from 32% in 1990 to 25% in 2018 (Funk and Parker, 2018). Diversifying the workforce will likely result in better-designed products, services and solutions that are more representative of the general public.

Due to gender discrimination, women in STEM face specific challenges such as higher demands from society to demonstrate more evidence or higher standards than their male counterparts. Most Asian women working in STEM fields report hearing from co-workers or family members that taking time off after having a child is a good idea. Women are often expected to take on stereotypically feminine roles at work while also doing care work at home (Jean, 2015; Tabassum, 2021). It is common for employed mothers to be said to work 'double shifts' because they continue to take on more housework and childcare than men. According to research by McKinsey, working mothers have been drastically affected by the pandemic in 2020 and 2021. Their challenges include managing household responsibilities, mental health challenges, the challenges of remote work and fears about higher rates of unemployment – especially among women of colour. Work-related burdens for working women are compounded by structural barriers such as being the 'only' female member of the team and playing an allyship role (Huang, 2021). This issue is not restricted to the Asian continent, and similar observations have been made in other parts of the world. A study by Suter (2006) of 600 Swiss university students demonstrates that women favour careers that provide useful skills for childrearing and do not conflict with family responsibilities, such as education, psychology or medicine. It would appear that women are less inclined to choose STEM fields due to their family responsibilities (OECD, 2008). Furthermore, Xie (2006) indicates that juggling family and work may be more difficult in some fields (for example, those requiring long lab hours) than in others (for example, social sciences). In contrast to men, women are more likely to be drawn to careers that allow them to interact with people as opposed to numbers, due to gendered forms of socialization (Betz and O'Connell, 1989; Baker and Leary, 1995; OECD, 2008; Ceci and Williams, 2011). Similarly, Gilbert et al (2003) point out that in Switzerland 'empirical evidence suggests that young men make their choice mostly based on career prospects, whereas women are also motivated by social and/or political commitments'. Research done by the OECD (2008) found that 'students who evaluate social skills and key competences as important for working in a modern economy may be discouraged from pursuing engineering studies, especially young women'. Stereotypes, cultural norms and gender roles also cause women to be segregated into certain fields of study. Suter (2006) also argues that stereotypes deter women from careers in STEM fields, since many believe such careers are in line with male rather than female characteristics. OECD (2008), Suter (2006) and Xie (2006) also noted that family background and a lack of female role models can influence female participation in STEM careers. Furthermore, Black and other women of colour who are employed in STEM report being mistaken for custodial and administrative staff, rather than being recognized for their professional expertise (Hammonds, 2021). Besides family pressure and feminine roles, the other form of discrimination/ harassment that women face in workplaces includes sexual harassment. According to a survey of US adults conducted in 2017, half of women working in STEM are likely to experience gender discrimination at work; about a fifth report having experienced sexual harassment themselves (Funk and Parker, 2018).

Developed by the World Economic Forum, the Global Gender Gap Index tracks progress towards closing gender-based gaps in four areas (economic participation and opportunity, education attainment, health and survival, and political empowerment) over time. As a result of the COVID-19 pandemic, as well as the associated economic downturn, women have been adversely affected, partially reopening previously closed gaps. Moreover, the report suggests that organizations and professional cultures in the area of computer science and technology may perpetuate gender segregation. Under-representation of women in STEM fields leads to stereotypical views of maths and science as masculine domains and the belief that men are superior in technical and maths–intensive fields (Ertl, 2017; Hand, 2017; Makarova, 2019). Due to the fact that gender stereotypes reinforce each other, there is a gender gap in career-related interests and choices, which, in turn, reinforces gender stereotypes. Women's inclusion in the workforce and addressing the gender gap in economic participation can be enabled by providing flexible or alternative work arrangements for diverse workforces, setting leadership goals at the business and government level, and providing childcare support.

The STEM gender gap needs to be closed at multiple levels. Discrimination against women in STEM majors and at work persists throughout education and the workplace. Providing a variety of STEM opportunities for children, including science fairs and coding camps, is essential for parents to encourage children that they can be anything they want. Scholarships and grants can be used to encourage women to pursue STEM fields as well. For retention to improve, however, it is imperative that these departments take every step possible to prevent sexism and discrimination against female students. In order to foster greater productivity and employee happiness, employers also need to systematically identify and address discriminatory hiring practices (Kong, 2020). According to the best practices identified in management–employee relationships and salary scales, men and women should also be treated equally. However, schools and the workplace are not the only places where societal changes are needed. Starting a family has a profound effect on women, forcing them to take time off during critical junctures in their careers, such as when they are starting a new job or studying for their doctorate.

Through a framework that promotes and transforms gender equality, women's participation and representation in STEM education and research can be accelerated. It is recognized that women are key to removing structural barriers, such as economic, cultural and social ones, including misogyny, patriarchy and embedded structures, that prevent women from achieving their full potential as part of a gender-equality framework. Such a framework can be employed by STEM institutions to conduct comprehensive self-assessments to identify barriers and norms specific to their departments, degrees or disciplines, and to develop interventions to overcome these. It can also be used to collect qualitative data to establish gender-equality benchmarks in order to determine whether barriers exist and how they might be eliminated. Qualitative criteria can be used for the retention of women in STEM careers in a variety of ways, ranging from creating support networks for career advancement to promoting inclusive workplace practices. By establishing a framework of this nature, institutions can further enhance gender equality while providing support to their funders and research councils at the national and international levels.

As reported in the latest All–India Survey on Higher Education (AISHE), the gender gap in India has narrowed over the past few years. Women now make up almost half of higher education students (48.6%). Women in STEM fields have benefited from government efforts to progress gender equality and gender advancement in academic and research institutions. The Government of India (the Science and Technology Department and University Grants Commission have recently begun to place a great deal of emphasis on the education of women and are making considerable efforts to improve women's skills. Students, and especially women, are encouraged to pursue careers in science and technology through a targeted scholarship and awards programmes. Indian women are climbing the academic and administrative ladders, despite many structural and systemic obstacles. While the enrolment ratio in India has improved over the years, a gender-equality framework for higher education institutions would be an important step forward (Keohane, 2020). Some research avenues are suggested by anecdotal membership data from AHIMA (American Health Information Management Association). Currently, 1.7% of credentialled female members hold executive or vice-president positions in healthcare organizations, compared to 4.3% of their male counterparts. Further, anecdotal membership data from the AHIMA survey indicated that not only is the number of opportunities less for women; the earnings of women also continue to lag behind those of men. All educational levels are affected by the pay gap, with women earning 77% percent less than their male counterparts when they work full time and year-round. Female high school graduates earned 69.6 cents for every dollar earned by their male counterparts, and female college graduates earned 70.9 cents for every dollar earned by their male counterparts. A woman's pay gap is most pronounced at the top of the educational spectrum, where she earns 57.9 cents for every dollar a man earns. According to the data reported in 2019, earnings of women working full-time, year-round in the United States were $48,096 in 2020 while it is $62,843 for men referring that women earned 76.3 cents for every dollar that men earned (Johns, 2013; ACE, 2021). As well as aligning with the ambitious goals of NEP 2020, this will also be crucial to achieving UN Sustainable Development Goal 5 (NEP 2020).

Gender bias in science communication

The under-representation of female researchers in STEM fields is widespread. One factor contributing to this may be how other scientists view their work (for example, National Science Foundation, 2006; European Commission, 2012). Specifically, the gender of an author as well as the topic of their research could determine how they are evaluated for science quality (Handley, 2015). Public, scientific, institutional, political and ethical views of change, in all their diversity and heterogeneity, need to be linked in new ways, argues Irwin (Irwin and Wynne, 1996). Broks also calls for the development of a new perspective known as 'Critical Understanding of Science in Public (CUSP)' (Broks, 2006). Based on this paradigm, science communication should make use of several approaches by taking into account social, political and cultural contexts, incorporating values and opinions from lay people as well as professional experts and focusing on meaning rather than content. Scientists and science communicators must change their mindsets as well as their institutional values if they are to embrace CUSP and third-order thinking. In addition to the deficit, dialogue and participatory science communication models, engagement activities must also consider full-spectrum complexity. Education, public health messaging science promotion within and between disciplines and news related to science, environment and medicine are all part of science communication. The study and practice of science communication is varied in terms of interests, objectives and approaches. Historically, marginalized groups (vis-à-vis race, ethnicity, culture, religion, gender, sexuality and socioeconomic status) are under-represented in the practice and study of science communication (Broks, 2006).

The term 'Matilda effect' – under-recognition of female scientists in science communication due to gender roles and responsibilities – was coined and studied by Knobloch-Westerwick (2013). Specifically, more feminine-stereotyped characteristics (for example, empathy, emotion, communication) are thought to contribute to career success in fields such as human resources and social work, while more masculine-stereotyped characteristics (objectivity, organization, reason) are thought to contribute to success in science and leadership fields. Those who are incongruent with their gender roles in this field will be unable to assess the quality of science and to collaborate effectively. The study presents otherwise identical scientific research to test the hypothesis that men's scholarly work has a higher scientific quality than women's work in the same field and that men are more likely to exchange scholarly ideas with other men. Despite being more egalitarian in nature, higher education does not reduce differences associated with gender ideology (Bryant, 2003). Men cite same-sex authors more prevalently than women, due to the Matilda effect, while women cite both male and female authors according to the amount of published material they have access to (Knobloch-Westerwick and Glynn, 2013). Also, research topics are different, based on whether they are usually associated with male or female stereotypes. Women who work in the communication field who are involved in political communication may show more incongruence between their gender roles and their scientific roles than women who work in the media or with children. It has been demonstrated empirically that such stereotyping of communication topics is present in the proportions of males and females among authors in communication research publications (Knobloch-Westerwick and Glynn, 2013; Kiprotich and Changorok, 2015; Lindvall-Östling, 2020).

It has also been shown that the female researchers received fewer grants, smaller grants (Wenners and Wold, 1997; RAND, 2005; Bornmann, 2007) and fewer citations (for example, Knobloch-Westerwick and Glynn, 2013) than their male counterparts. For example, 'men were more than eight times more likely than women to win a scholarly award and almost three times more likely to win a young investigator award' (Lincoln, 2012). Despite this shocking under-representation, its causes remain a matter of debate (for example, Ceci, 2009). Science communication, as part of communication among scholars, may be one of the contributing factors. In addition to scholarly communication among scientists, several conceptualizations of science communication have been proposed. For example, Burns, O'Connor, and Stocklmayer (2003) note that 'Science communication may involve science practitioners, mediators, and other members of the general public, either peer-to-peer or between groups.' Citation analysis has been used to study scholarly communication (for example, Tai, 2009; Knobloch-Westerwick and Glynn, 2013), which shows how one scholar's work is utilized (that is, cited) by another. Most commonly, academic communication is viewed as a one-way exchange, with the recipient merely receiving and perceiving another scientist's publication. While it is assumed that scientists communicate objectively in an ideal market of ideas, there are always patterns of stereotyping and relations of power (for example, Greenwald and Banaji, 1995). Scientists' communication can be seen as a social system (for example, Garvey and Griffith, 1967) and therefore is subject to bias (for example, Garvey and Griffith, 1967).

Conclusion

Social criticism plays a crucial role in science, technology and ethics. As social change occurs, social commentators seek to explain and guide it, while also challenging its assumptions and directions. Social theories often include assessments of the social world (positive or negative). In the mid-19th century, Karl Marx (1818–83) saw the devastating effects of rapid industrialization and urbanization in Manchester, England as the means of production determined the structure of society and belief systems. On the other hand, a wide range of issues concerning scholars of the 21st century, like explaining new technologies and their effects on Indigenous cultures, have come to the conclusion that societies are threatened by technological advancements (Mukhtar, 2013; Anderson and Raine, 2018). When we focus on how technology affects work and employment, we are often concerned with problems of inequality and social stratification. Social theory intersects with science and technology on the political and ethical levels, since it deals with human dimensions and both the causes and consequences of change. Within feminist STS, multiple voices interrogate, displace and rethink subjectivity, to the point that we were drawn to a better understanding of the proliferating paths of research and the increasing effects of interventions made by feminist concerns.

For systematic knowledge generation and for appreciating science as a key factor of progress, every country should take the following actions: provide access to current scientific information; align national scientific goals with scientific research conducted in the country; foster a social interest in science and its applications; develop scientific resources in multiple ways; and enhance scientific literacy among the general public (Gutiérrez-Vargas, 2002). It is critical that the development of technology associated with our health should reflect this understanding, and take into account that all areas of our lives are interconnected – family, work, finances and social networks. And when it comes to providing holistic care solutions, this must be recognized.

Science communication and public engagement research in social science strongly suggests that audiences must transcend narrow notions of audience and the deficit in effective science communication and public engagement. Researchers and practitioners in science communication need to prioritize inclusion, equity and intersectionality in order to engage and benefit the entire community with science communication. During the 21st century, health communication, that is, interpersonal and mass communication activities that aim to improve public health, has become increasingly important. The concept of health literacy, or the ability to understand what constitutes health, refers to the ability to comprehend and apply health information, which, in turn, impacts on health behaviours and outcomes. Over the past few decades, feminist scholars and sociologists of science have investigated the gendered nature of science and its representation. Insofar as science is a human activity, patriarchy and gender bias are present, especially in its interpretation and representation.

References

Abouelmehdi, K., Beni-Hessane, A. and Khaloufi, H. (2018) 'Big healthcare data: preserving security and privacy', *Journal of Big Data*, 5(1), https://doi.org/10.1186/s40537-017-0110-7.

Adrian, S.W. (2014) 'Assisteret befrugtning, en feministisk teoretisk udfordring?' *Kvinder, Køn & Forskning*, 23(3): 54–68.

Almathami, H. K. Y., Win, K. T. and Vlahu-Gjorgievska, E. (2020) 'Barriers and facilitators that influence telemedicine-based, real-time, online consultation at patients' homes: systematic literature review', *Journal of Medical Internet Research*, 22(2): e16407, https://doi.org/10.2196/16407.

Anderson, J. and Raine, L. (2018) 'The future of well-being in a digital world', Pew Research Center, https://assets.pewresearch.org/wp-content/uploads/sites/14/2018/04/14154552/PI_2018.04.17_Future-of-Well-Being_FINAL.pdf.

Baciu, A. (2017) *Health Equity and Nursing: Achieving Equity through Policy, Practice, and Research*, New York: Springer.

Baker, D. and Leary, R. (1995) 'Letting girls speak out about science', *Journal of Research in Science Teaching*, 32(1): 3–27.

Barad, K. (1995) 'A feminist approach to teaching quantum physics', in S.V. Rosser (ed) *Teaching the Majority: Breaking the Gender Barrier in Science, Mathematics, and Engineering*, New York: Athene Series, Teacher's College Press, pp 43–75.

Barad, K. (2007) *Meeting the Universe Halfway: Quantum Physics and the Entanglement of Matter and Meaning*, Durham, NC: Duke University Press.

Bendelow, G. (1993) 'Pain perceptions, emotions and gender', *Sociology of Health & Illness*, 15(3): 273–294.

Betz, M. and O'Connell, L. (1989) 'Work orientations of males and females: exploring the gender socialization approach', *Sociological Inquiry*, 59(3): 318–330, https://doi.org/10.1111/j.1475-682x.1989.tb00109.x.

Birke, L. (1986) *Women, Feminism and Biology: The Feminist Challenge*, Brighton: Wheatsheaf.

Bornmann, L., Mutz, R. and Daniel, H.D. (2007) 'Gender differences in grant peer review: a meta-analysis', *Journal of Infometrics*, 1: 226–238, https://doi.org/10.1016/ j.joi.2007.03.001.

Boyd, D. and Crawford, K. (2012) 'Critical questions for Big Data', *Information, Communication & Society*, 15(5): 662–679, https://doi.org/10.1080/1369118X.2012.678878.

Broks, P. (2006) *Understanding Popular Science*, Milton Keynes: The Open University Press.

Bryant, A.N. (2003) 'Changes in attitudes toward women's roles: predicting gender-role traditionalism among college students', *Sex Roles*, 48: 131–142, https://doi.org/10.1023/A:1022451205292.

Burns, T.W., O'Connor, D.J. and Stocklmayer, S.M. (2003) 'Science communication: a contemporary definition', *Public Understanding of Science*, 12: 183–202, https://doi.org/10.1177/09636625030122004.

Cardona-Morrell, M., Kim, J.C.H., Turner, R.M., Anstey, M., Mitchell, I.A. and Hillman, K. (2016) 'Non-beneficial treatments in hospital at the end of life – a systematic review on extent of the problem', *International Journal for Quality in Health Care*, 28(4): 456–469.

Carrera, P.M., Kantarjiam, H.M. and Blinder, V.S. (2018) 'The financial burden and distress of patients with cancer – understanding and stepping-up action on the financial toxicity of cancer treatment', *CA: A Cancer Journal for Clinicians*, 68(2): 153–165.

Cavoukian, A. and Emam, K. (2014) *De-identification Protocols – Essential for Protecting Privacy*, Privacy by Design, 25 June, www.privacybydesign.ca/content/uploads/2014/06/pbd-deidentifcation_essential.pdf.

Ceci, S.J. and Williams, W.M. (2011) 'Understanding current causes of women's underrepresentation in science', *Proceedings of the National Academy of Sciences*, 108(8): 3157–3162.

Ceci, S J., Williams, W.M. and Barnett, S.M. (2009) 'Women's underrepresentation in science: sociocultural and biological considerations', *Psychological Bulletin,* 135: 218–261, https://doi.org/10.1037/a0014412.

Chan, P.C., Handler, T. and Frenette, M. (2021) 'Gender differences in STEM enrolment and graduation: what are the roles of academic performance and preparation?' *Economic and Social Reports*: 1(11): 1–21.

Cirillo, D., Catuara-Solarz, S., Morey, C., Guney, E., Subirats, L., Mellino, S. et al (2020) 'Sex and gender differences and biases in artificial intelligence for biomedicine and healthcare', *npj Digital Medicine*, 3: 81, https://doi.org/10.1038/s41746-020-0288-5.

Claudia, P. (2020) 'Health communication practice strategies and theories', in C.F. Parvanta and S.B. Bass (eds) *Health Communication: Strategies and Skills for a New Era*, Burlington, MA: Jones and Bartlett Learning, pp 69–84.

Clifton, C. and Tassa, T. (2013) 'On syntactic anonymity and differential privacy', *Transactions on Data Privacy,* 6(2): 161–183.

Croucher, S.M. and Cronn–Mills, D. (2019) 'Ethnography', in *Understanding Communication Research Methods, A Theoretical and Practical Approach* (2nd edn), New York and London: Routledge, pp 85–100.

DeBoer, G.E. (2000) 'Scientific literacy: another look at its historical and contemporary meanings and its relationship to science education reform', *Journal of Research in Science Teaching,* 37(585), https://doi.org/10.1002/1098- 2736(200008)37:63.0.CO;2-L.

Devaraj, S.J. (2019) 'Emerging paradigms in transform-based medical image compression for telemedicine environment', in H.D. Jude and V.E. Balas (eds) *Telemedicine Technologies: Big Data, Deep Learning, Robotics, Mobile and Remote Applications for Global Healthcare*, New York: Academic Press, Elsevier, pp 15–30.

Dunwoody, S. (2014) 'Science journalism: prospects in the digital age', in M. Bucchi and B. Trench (eds) *Routledge Handbook of Public Communication of Science and Technology* (2nd edn), New York: Routledge, pp 27–39.

Ertl, B., Luttenberger, S. and Paechter, M. (2017) 'The impact of gender stereotypes on the self-concept of female students in stem subjects with an under-representation of females', *Frontiers of Psychology*, 8: 703, https://doi.org/10.3389/fpsyg.2017.00703.

Ernst, W. and Horwath, I. (eds) (2014) *Gender in Science and Technology: Interdisciplinary Approaches*, Bielefeld: Transcript Verlag.

European Commission (2012) Meta-analysis of gender and science research, www.genderandscience.org/doc/synthesis_report.pdf.

Eysenbach, G. (2008) 'What is eHealth?', *Journal of Medical Internet Research*, 10(Suppl 1): e20, https://doi.org/10.2196/jmir.10.1.e20.

Fausto-Sterling, A. (1985) *Myths of Gender: Biological Theories about Women and Men*, New York: Basic Books.

Fausto-Sterling, A. (2012) *Sex/Gender: Biology in a Social World*, New York: Routledge.

Fried, C. (1968) 'Privacy', *Yale Law Journal*, 77: 475–493.

Funk, C. and Parker, K. (2018) 'Women and men in STEM often at odds over workplace equity', Pew Research Center.

Ganasegeran, K., and Abdulrahman, S. (2019) 'Adopting m-health in clinical practice – a boon or a bane?' in H.D. Jude and V.E. Balas (eds) *Telemedicine Technologies: Big Data, Deep Learning, Robotics, Mobile and Remote Applications for Global Healthcare*, New York: Academic Press, Elsevier, pp 31–41.

Garvey, W.D. and Griffith, B.C. (1967) 'Scientific communication as a social system', *Science*, 157: 1011–1016, https://doi.org/10.1126/science.157.3792.1011.

genSET (2011) 'Advancing excellence in science through gender equality', Document prepared for the genSET Capacity Building Workshop 'Advancing Excellence in Science through Gender Equality', 28–29 March.

Gilbert, F., Crettaz Roten, F. and E. Alvarez, E. (2003) *Promotion des femmes dans les formations supérieures techniques et scientifiques. Rapport de recherche et recommandations,* Lausanne: Observatoire EPFL Science, Politique et Société.

Govender, V. and Penn-Kekana, L. (2007) 'Gender Biases and Discrimination – A Review of Health care Interpersonal Interactions', Background paper prepared for the Women and Gender Equity Knowledge Network of the WHO Commission on Social Determinants of Health, www.who.int/social_determinants/resources/gender_biases_and_discrimination_wgkn_2007.pdf.

Greenwald, A.G. and Banaji, M.R. (1995) 'Implicit social cognition: attitudes, self-esteem, and stereotypes', *Psychological Review*, 102: 4–27, https://doi.org/10.1037/0033-295X.102.1.4.

GSMA (2020) 'Connected women – the mobile gender gap report 2020', London: GSMA, www.gsma.com/mobilefordevelopment/wp-content/uploads/2020/05/GSMA-The-Mobile-Gender-Gap-Report-2020.pdf.

Guide to Privacy and Security of Electronic Health Information v2.0 (2015) Office of the National Coordinator for Health Information Technology, www.healthit.gov/sites/default/files/pdf/privacy/privacy-and-security-guide.pdf.

Gutiérrez-Vargas, J. (2002) 'Science and technology for sustainable development: the role of science education', *International Journal of Science Education*, 24(10): 1201–1216.

Haleem, A., Javaid, M., Singh, R.P. and Suman, R. (2021) 'Telemedicine for healthcare – capabilities, features, barriers, and applications', *Sensors International*, 2: 100117.

Hammonds, E., Taylor, V. and Hutton, R. (2021) *Transforming Trajectories for Women of Color in Tech*, Washington, DC: The National Academies Press.

Hand, S., Rice, L. and Greenlee, E. (2017) 'Exploring teachers' and students' gender role bias and students' confidence in STEM fields', *Social Psychology of Education,* 20: 929–945, https://doi.org/10.1007/s11218-017-9408-8.

Handley, I.M., Brown, E.R., Moss-Racusin, C.A. and Smith, J.L. (2015) 'Quality of evidence revealing subtle gender biases in science is in the eye of the beholder', *Proceedings of the National Academy of Sciences*, 112(43): 13201, https://doi.org/10.1073/pnas.1510649112.

Haraway, D. (1987) *Primate Visions: Gender, Race, and Nature in the World of Modern Science*, New York: Routledge.

Haraway, D. (1988) 'Situated knowledges: the science question in feminism and the privilege of partial perspective', *Feminist Studies*, 14(3): 575–599.

Harding, S. (1986) *The Science Question in Feminism*, Ithaca: Cornell University Press.

Helms, R.M., Schendel, R., Godwin, K. and Blanco, D. (eds) (2021) *Women's Representation in Higher Education Leadership Around the World*, Boston College, Center for International Higher Education, www.ace net.edu/Documents/Womens-Rep-in-Higher-Ed-Leadership-Around-the-World.pdf.

Horgan, D., Metspalu, A., Ouilade, M., Athanasiou, D., Pasi, J., Adjali, O. et al (2020) 'Propelling healthcare with advanced therapy medicinal products – a policy discussion', *Biomed Hub*, 5: 511678.

Huang, J., Krivkovich, A., Rambachan, I. and Yee, L. (2021) *For Mothers in the Workplace, a Year (and Counting) Like No Other*, London: McKinsey and Company.

Institute of Medicine (US) Committee on Health Research and the Privacy of Health Information (2009) 'The HIPAA Privacy Rule', in S.J. Nass, L.A. Levit and L.O. Gostin (eds) *Beyond the HIPAA Privacy Rule: Enhancing Privacy, Improving Health Through Research*, Washington, DC: National Academies Press.

Irwin, A. and Wynne, B. (eds) (1996) *Misunderstanding Science? The Public Reconstruction of Science and Technology*, Cambridge: Cambridge University Press.

Jean, V.A., Payne, S.C. and Thompson, R.J. (2015) 'Women in STEM: family-related challenges and initiatives', in M. Mills (ed) *Gender and the Work–Family Experience: An Intersection of Two Domains*, New York: Springer, pp 291–311.

Jogova, M., Shaw, J. and Jamieson, T. (2019) 'The regulatory challenge of mobile health: lessons for Canada', *HealthCare Policy*, 14(3): 19–28.

Johns, M.L. (2013) 'Breaking the glass ceiling: structural, cultural, and organizational barriers preventing women from achieving senior and executive positions', *Perspectives in Health Information Management* 2013(10): 1e.

Keller, E.F. (1992) *Secrets of Life, Secrets of Death: Essays on Language, Gender and Science*, London: Routledge.

Keohane, N.O. (2020) 'Women, power and leadership', *American Academy of Arts & Sciences*, 149(1): 236–250.

Kiprotich, A.J. and Changorok, J.R. (2015) 'Gender communication stereotypes: a depiction of the mass media', *IOSR Journal of Humanities and Social Science (IOSR-JHSS)*, 20(11): 69–77.

Knobloch-Westerwick, S. and Glynn, C.J. (2013) 'The Matilda effect – role congruity effects on scholarly communication: a citation analysis of *Communication Research* and *Journal of Communication* articles', *Communication Research*, 40(1): 3–26, https://doi.org/10.1177/0093650211418339.

Knowles, B. and Hanson, V.L. (2018) 'The wisdom of older technology (non) users', *Communications of the ACM*, 61(3): 72, https://doi.org/10.1145/3179995.

Kong, S.M., Carroll, K.M., Lundberg, D.J., Omura, P. and Lepe, B.A. (2020) 'Reducing gender bias in STEM', *MIT Science Policy Review*, 1: 55–63.

Kory-Westlund, J.M. and Breazeal, C. (2019) 'A persona-based neural conversation model', in J.A. Fails (ed) *Proceedings of the 18th ACM Interaction Design and Children Conference (IDC)*, Boise, ID: ACM Press, pp 38–50.

Kruk, M.E., Gage, A.D., Arsenault, C., Jordan, K., Leslie, H.H., Roder-DeWan, S. et al (2018) 'High-quality health systems in the Sustainable Development Goals era – time for a revolution', *The Lancet Global Health Commission*, 6: e1196–e1252.

Kruse, C., Betancourt, J., Ortiz, S., Luna, S.M.V., Bamrah, I. and Segovia, N. (2019) 'Barriers to the use of mobile health in improving health outcomes in developing countries: systematic review', *Journal of Medical Internet Research*, 21(10): e13263.

Ku, J.P. and Sim, I. (2021) 'Mobile health: making the leap to research and clinics', *Digital Medicine,* 4: 83, https://doi.org/10.1038/s41746-021-00454-z.

Laugksch, R.C. (1999) 'Scientific literacy: a conceptual overview', *Science Education*, 84(1): 71–94, https://doi.org/10.1002/(SICI)1098-237X(200001)84:1<71::AID-SCE6>3.0.CO;2-C.

Law, J. (2017) 'STS as method', in U. Felt, R. Fouche, C. Miller and L. Smith-Doerr (eds) *Handbook of Science and Technology Studies* (4th edn), Boston, MA: MIT Press, pp 31–57.

Lincoln, A.E., Pincus, S., Koster, J.B. and Leboy, P.S. (2012) 'The Matilda effect in science: awards and prizes in the US, 1990s and 2000s', *Social Studies of Science,* 42: 307–320, https://doi.org/10.1177/0306312711435830.

Lindvall-Östling, M., Deutschmann, M. and Steinvall, A. (2020) 'An exploratory study on linguistic gender stereotypes and their effects on perception', *Open Linguistics*, 6(1): 567–583, https://doi.org/10.1515/opli-2020-0033.

Litwin, A.S. (2020) *Technological Change in Health Care Delivery: Its Drivers and Consequences for Work and Workers*, Report from the UC Berkeley Center for Labor Research and Education and Working Partnerships, USA.

Logan, D.E., Breazeal, C., Goodwin, M.S., Jeong, S., O'Connell, B., Smith-Freedman, D. et al (2019) 'Social robots for hospitalized children', *Pediatrics*, 144: e20181511.

Lupton, D. (2014). *The Digitally Divided Self: The Social, Ethical and Political Implications of the Quantified Self*, New York: Routledge.

Lykke, N. (1996) 'Between, monsters, goddesses and cyborgs: feminist confrontations with science', in N. Lykke and R. Braidotti (eds) *Monsters, Goddesses and Cyborgs: Feminist Confrontations with Science, Medicine and Cyberspace*, London: Zed Books, pp 13–29.

Makarova, E., Aeschlimann, B. and Herzog, W. (2019) 'The gender gap in STEM fields: the impact of the gender stereotype of math and science on secondary students' career aspirations', *Frontiers in Education*, 4: 60, https://doi.org/10.3389/feduc.2019.00060.

Malvey, D.M. and Slovensky, D.J. (2014) *mHealth: Transforming Healthcare*, New York: Springer.

Margolis, J. and Fisher, A. (2002) *Unlocking the Clubhouse: Women in Computing*, Cambridge, MA: MIT Press.

Mathews, S.C., McShea, M.J., Hanley, C.L., Ravitz, A., Labrique, A.B. and Cohen, A.B. (2019) 'Digital health – a path to validation', *npj Digital Medicine*, 2: 38.

Ministry of Human Resource Development (2020) National Education Policy 2020, New Delhi: Government of India.

Moore, A. (2005) 'Intangible property: privacy, power and information control', in A. Moore (ed) *Information Ethics: Privacy, Property, and Power*, Seattle, WA: University of Washington Press, pp 101–114.

Mukhtar, S. (2013) *About the Impact of Technology upon Society*, Munich: GRIN Verlag, www.grin.com/document/205584.

Murphy, M. (2012) *Seizing the Means of Reproduction: Entanglements of Feminism, Health and Technoscience*, Durham, NC: Duke University Press.

National Science Foundation (2006) *Beyond Bias and Barriers: Fulfilling the Potential of Women in Academic Science and Engineering*, Arlington, VA: National Science Foundation.

Neumann, P.J. and Weinstein, M.C. (1991) 'The diffusion of new technology: costs and benefits to health', in A.C. Gelijns and E.A. Halm (eds) *The Changing Economics of Medical Technology*, Washington, DC: National Academies Press.

NRC (2007) *Engaging Privacy and Information Technology in a Digital Age*, Washington, DC: National Academies Press.

OECD (2008) *Encouraging Student Interest in Science and Technology Studies*, Paris: OECD Publishing.

OECD (2012) *Measuring Regulatory Performance – Evaluating the Impact of Regulation and Regulatory Policy*, Paris: OECD Publishing.

OECD (2013) *Recommendation of the Council concerning Guidelines Governing the Protection of Privacy and Transborder Flows of Personal Data*, amended on 11 July, Paris: OECD, https://legalinstruments.oecd.org/public/doc/114/114.en.pdf.

OECD (2013a) *Strengthening Health Information Infrastructure for Health Care Quality Governance: Good Practices, New Opportunities and Data Privacy Protection Challenges*, Paris: OECD Health Policy Studies.

OECD (2013b) The OECD Privacy Framework, www.oecd.org/sti/iecon omy/oecd_privacy_framework.pdf.

OECD (2015) *Health Data Governance: Privacy, Monitoring and Research*, Paris: OECD Publishing, https://read.oecd-ilibrary.org/social-issues-migration-health/health-data-governance_9789264244566-en#page47.

OCED (2018) *Report: Bridging the Digital Gender Divide – Include, Upskill, Innovate*, www.oecd.org/digital/bridging-the-digital-gender-divide.pdf.

OECD (2020) *Mapping Approaches to Data and Data Flows*, Saudi Arabia: OECD Publishing, www.oecd.org/sti/mapping-approaches-to-data-and-data-flows.pdf.

OHRP (2008) 'Guidance on Research Involving Private Information or Biological Specimens', Department of Health and Human Services, Office of Human Research Protections (OHRP), 16 August, www.hhs.gov/ohrp/policy/cdebiol.html.

Oudshoorn, N. (1994) *Beyond the Natural Body: An Archaeology of Sex Hormones*, London: Routledge.

Parvanta, C. (2020) 'Science communication in the age of COVID-19', *Journal of Science Communication*, 19(2): A02, https://doi.org/10.22323/2.1902.A02.

Ploug, T. (2020) 'In defence of informed consent for health record research – why arguments from "easy rescue", "no harm" and "consent bias" fail', *BMC Medical Ethics*, 21: 75, https://doi.org/10.1186/s12910-020-00519-w.

Pritts, J. (2008) 'The importance and value of protecting the privacy of health information: roles of HIPAA Privacy Rule and the Common Rule in health research', http://www .iom.edu/CMS/3740/43729/53160 .aspx.

PTAC (Privacy Technical Assistance Center), US Department of Education (2013) '*Data de-identification: an overview of basic terms*', https://studentpriv acy.ed.gov/sites/default/files/resource_document/file/data_deidentific ation_terms.pdf.

RAND (2005) 'Is there gender bias in federal grant programs?' RAND Infrastructure, Safety, and Environment Research Brief No. RB-9147-NSF, http://rand.org/pubs/research_briefs/RB9147/RAND_RB9147.pdf.

Rapp, R. (2000) *Testing Women, Testing the Fetus: The Social Impact of Amniocentesis in America*, New York: Routledge.

Regitz-Zagrosek, V. (2012) 'Sex and gender differences in health – Science & Society series on Sex and Science', *EMBO Reports*, 13(7): 596–603.

Rivenbark, J.G. and Ichou, M. (2020) 'Discrimination in healthcare as a barrier to care: experiences of socially disadvantaged populations in France from a nationally representative survey', *BMC Public Health*, 20: 31, https://doi.org/10.1186/s12889-019-8124-z.

Samulowitz, A., Gremyr, I., Eriksson, E. and Hensing, G. (2018) '"Brave men" and "emotional women": a theory-guided literature review on gender bias in health care and gendered norms towards patients with chronic pain', *Pain Research and Management*, 2018: 6358624, https://doi.org/10.1155/2018/6358624.

Shin, D. and Choi, M.J. (2015) 'Ecological views of Big Data: perspectives and issues', *Telematics and Informatics*, 32(2): 311–320.

Singleton, P. and Wadsworth, M. (2006) 'Consent for the use of personal medical data in research', *BMJ*, 333(7561): 255–258, https://doi.org/10.1136/bmj.333.7561.255.

Star, S.L. (1991) 'Power, technology and the phenomenology of conventions: on being allergic to onions', in J. Law (ed) *A Sociology of Monsters: Essays on Power, Technology and Domination*, London: Routledge, pp 26–56.

Stewart-Williams, S. and Halsey, L.G. (2021) 'Men, women and STEM: why the differences and what should be done?' *European Journal of Personality*, 35(1): 3–39.

Suter, C. (2006) '"Trends in Gender Segregation by Field of Work in Higher Education." Institut de Sociologie, University of Neuchatel Switzerland', in *OECD, Women in Scientific Careers: Unleashing the Potential*, Paris: OECD.

Tabassum, N. and Nayak, B.S. (2021) 'Gender stereotypes and their impact on women's career progressions from a managerial perspective', *IIM Kozhikode Society & Management Review*, 10(2): 192–208, https://doi.org/10.1177/2277975220975513.

Tai, Z. (2009) 'The structure of knowledge and dynamics of scholarly communication in agenda setting research, 1996–2005', *Journal of Communication*, 59: 481–513, https://doi.org/10.1111/j.1460-2466.2009.01425.x.

Tarango, J. and Machin-Mastromatteo, J.D. (2017) *The New Profile of Information Professionals as Scientific Production and Communication Managers: Identification of Competences in the Role of Information Professionals in the Knowledge Economy: Skills, Profile and a Model for Supporting Scientific Production and Communication*, Cambridge, MA: Chandos Publishing/Elsevier.

Taylor, K. (2015) *Connected Health: How Digital Technology is Transforming Health and Social Care*, Deloitte Centre for Health Solutions, London: Deloitte LLP.

Terry, N.P. and Francis, L.P. (2007) 'Ensuring the privacy and confidentiality of electronic health records', *University of Illinois Law Review*, 2007(2): 681–736.

Trojer, L. (1985) 'Kvinnoperspektiv på naturvetenskapen', in M. Bryld and N. Lykke (eds) *Kvindespor i videnskaben*, Odense: Odense Universitetsforlag, pp 93–99.

UNESCO (2017) 'Digital society: gaps and challenges for digital inclusion in Latin America and the Caribbean', Paris: UNESCO, https://unesdoc. unesco.org/ark:/48223/pf0000262860_eng.

UNCTAD (2021) Technology and Innovation Report 2021: Catching Technological Waves: Innovation with Equity. New York and Geneva: UNCTAD.

Vlassoff, C. (2007) 'Gender differences in determinants and consequences of health and illness', *Journal of Health, Population and Nutrition,* 25(1): 47–61.

Vaportzis, E., Clausen, M.G. and Gow, A.J. (2017) 'Older adults' perceptions of technology and barriers to interacting with tablet computers: a focus group study', *Frontiers in Psychology*, 8: 1687, https://doi.org/10.3389/ fpsyg.2017.01687.

Vooren, M., Haelermans, C., Groot, W. and Brink, H.M. (2022) 'Comparing success of female students to their male counterparts in the stem fields: an empirical analysis from enrolment until graduation using longitudinal register data', *International Journal of STEM Education*, 9(1): 1–17.

Wajcman, J. (2004) *Technofeminism*, Cambridge: Polity Press.

Wajcman, J. (2006) 'Technocapitalism meets technofeminism: women and technology in a wireless world', *Labour & Industry: A Journal of the Social and Economic Relations of Work*, 16(3): 7–20, https://doi.org/10.1080/ 10301763.2006.10669327.

Wallis, L., Hasselberg, M., Barkman, C., Bogoch, I., Broomhead, S., Dumont, G. et al (2017) 'A roadmap for the implementation of mHealth innovations for image-based diagnositic support in clinical and public-health settings – a focus on front-line health workers and health-system organisations', *Global Health Action*, 10: 1340254.

Wamble, D.E., Ciarametaro, M. and Dubois, R. (2019) 'The effect of medical technology innovations on patient outcomes, 1990–2015: results of a physician survey', *Journal of Managed Care & Special Pharmacy*, 25(1): 66–71.

Wennerås, C. and Wold, A. (1997) 'Nepotism and sexism in peer-review', *Nature*, 387: 341–343, https://doi.org/10.1038/387341a0.

WHO (2001) *Health Care Technology Management: Health Care Technology Policy Framework*, Regional Office for the Eastern Mediterranean, https:// apps.who.int/iris/handle/10665/119642.

WHO (2008) *Health Technology Assessment and Health Policy-Making in Europe – Current Status, Challenges and Potential*, European Observatory on Health Systems and Policies, Garrido, M.V., Kristensen, F.B., Nielsen, C.P. and Busse, R., Geneva: World Health Organization.

WHO (2021) *Global Strategy on Digital Health 2020–2025*, www.who.int/docs/default-source/documents/gs4dhdaa2a9f352b0445bafbc79ca799dce4d.pdf.

Woodward, M. (2019) 'Cardiovascular disease and the female disadvantage', *International Journal of Environmental Research and Public Health*, 16(7): 1165, https://doi.org/10.3390/ijerph16071165.

World Social Report (2020) 'Inequality in a rapidly changing world', United Nations Department of Economic and Social Affairs, pp 1–256.

Wyatt, S., Milojevic, S., Park, H.W. and Leydesdorff, L. (2017) 'Quantitative and qualitative STS: the intellectual and practical contributions of scientometrics', in U. Felt, R. Fouche, C. Miller and L. Smith-Doerr (eds) *Handbook of Science and Technology Studies* (4th edn), Boston, MA: MIT Press, pp 87–102.

Xie, Y. (2006) '"Theories into Gender Segregation in Scientific Careers", Department of Sociology, University of Michigan, Ann Arbor, United States', in *Women in Scientific Careers: Unleashing the Potential*, Paris: OECD.

Yanyan, L. and Mengmeng, G. (2021) 'Scientific literacy in communicating science and socio-scientific issues: prospects and challenges', *Frontiers in Psychology*, 12, https://doi.org/10.3389/fpsyg.2021.758000.

Zara, M.C. and Monteiro, L.H.A. (2021) 'The negative impact of technological advancements on metal health: an epidemiological approach', *Applied Mathematics and Computation*, 396: 125905.

7

Automation in Medical Imaging: Who Gets What AI Sees? Insights from the Adopters' Perspective

Filomena Berardi and Giorgio Vernoni

Introduction

The impact of digital technologies on labour has become in the past decade a focus of science and technologies studies (STS). The potential effects that a group of integrated technologies – connectivity, Big Data, the increase in computing capacity, forms of machine learning – could have on the quantity of employment and on the quality of working practices in most sectors generate widespread fears and, at times, hopes for profound social change. The rationale used to estimate these effects, in the first stage of research, was primarily based on utilitarian techno-economic categories, that is, oriented to assess (a) the technical feasibility of automation and (b) the cost and, therefore, any competitive advantage that could derive from implementation. Less attention was paid to other important non-techno-economic factors – sociotechnical, cognitive-behavioural, historical and cultural – that constitute the core of the interdisciplinary approach of STS.

In this context, the application of digital technologies and artificial intelligence to medical imaging constitutes an ideal case for this dominant cognitive approach. In a sector that has long been deeply invested in the use of digital technologies, first in preclinical research and then in the therapeutic field, medical imaging has been considered an activity inevitably destined to be improved by new technologies, in terms of the quantity and quality of diagnostic activities, while clinicians and technicians were seen as predestined to be sacrificed on the altar of progress.

Is this accurate? The first experiences of Computer Aided Diagnosis systems from the 1990s onwards showed that the unfolding of these implications is much more complex and multifaceted than the reductionism of rational economic models can represent. The first experiences based on the use of machine-learning systems in work processes that seem ideal for their application have led to interesting but limited results and, above all, these have not become fully institutionalized in healthcare systems.

This chapter intends to exemplify some non-techno-economic factors that could be at the basis of these contradictory results, in order to articulate analytical tools that can be used to study the effects of new technologies on work in healthcare and in other sectors. Leaving aside factors connected to the legal responsibility of medical staff in carrying out diagnostic and therapeutic activity, which would require a dedicated discussion, the chapter analyses some factors that could discourage or condition the adoption of automation in medical imaging, paying particular attention to the relationships among clinicians and between clinicians and patients. To this end, a review of the existing literature was conducted, supported by a set of interviews with clinical radiologists in different organizational roles.

The chapter begins with a section on the current analytical framework adopted to analyse the impact of automation on work, highlighting its mainly techno-economic and utilitarian matrix. The next section contains an examination of the main applications of digital technologies in medical imaging and highlights the high theoretical risk of automation, at least according to the current analytical framework. A section then follows that tests this assumption, starting from the contradictory results of experiences based on Computer Aided Diagnosis systems, and tries to contextualize the implementation of these technologies in the light of a framework (Non-adoption, Abandonment, Scale- up, Spread, and Sustainability – NASSS) which assumes non-adoption as a possible outcome of the implementation process. The final section examines some hindering factors that could be at the basis of non-adoption from the perspective of medical staff, patients and the relationships that exist between them.

Some stylized facts on the automation of labour

Labour automation has been a key topic of socioeconomic research since 2010 and has gained relevance as a potential social risk factor. Automation, however, is not a recent concern; it can be traced back to as early as 1930, when John Maynard Keynes introduced the concept of 'technological unemployment' in relation to the loss of jobs caused by the mechanization of industrial activities (Keynes, 1910). Moving to the end of the 20th century, the debate on the relationship between technology and employment was dominated by the search for a recalibration of labour markets that favoured of employees with complementary skills with those required by technology (skill biased technical change) or, rather, by the detrimental effects of technology on easily automatable tasks (routine biased technical change). Instead, today's concern seems to be fuelled by the potentially disruptive effects of digital technologies, such as connectivity, the accumulation of large volumes of data and different forms of 'artificial intelligence' integrated with the latest generation of robotic and computer hardware (high performance computing) (Vernoni, 2018). In fact, these technologies allow for the replication of an increasingly wide array of not only manual but also cognitive tasks and activities. As a result, the potential for job automation is moving up the occupational system, affecting jobs conventionally requiring high skills and, in some cases, profiles that are really sophisticated, at least from a human perspective.

The first assessments of the possible impact of digital technologies on employment were carried out by adopting simulation models to estimate the risk of substitution based on a criterion of 'technical feasibility of automation'. Departing from standardized job description directories as a reference (in particular the US Occupational Information Network – O*NET), a degree of automatability of tasks and activities (that is, the performance of several tasks within a work process) that substantiate job profiles was assessed. Thus a hypothetical 'probability of automation' as a function of automatability was established and applied to different labour force data (Frey and Osborne, 2013; Arntz, Gregory and Zierahn, 2016; Nedekoska and Quintini, 2018). However, it should be emphasized that technical feasibility represents only one of the factors that, at the microeconomic level, can contribute to automation processes and on its own is not sufficient to determine an effective substitution effect.

Further studies have shown that other factors, such as cost, must be considered in estimating the risk of robotization. First, the automated production process must be economically advantageous as compared to one based on human labour (thus the incentive to automate changes in relation to labour cost, investment capacity, interest rates and so on) (Eurofound, 2019). Additional studies have focused on the way in which work is performed and coordinated, pointing out that high levels of polyvalence (that is, the number of tasks and activities that need to be performed in each job) and organizational coordination can limit the propensity for automation (Autor, 2015). Others have focused on the characteristics of employers, showing greater obstacles to automation for organizations with limited labour division and specialization (Arntz, Gregory and Zierahn, 2016). In short, it can be affirmed that the implementation of automation is subject to several conditions.

However, the clear prevalence of a techno-economic approach and the pronounced utilitarian orientation of these studies has led to an underestimation of significant cultural, behavioural and socio-organizational factors (Nikishawa and Bae, 2018). For example, little attention has been paid to the propensities and aversions of the various stakeholders in automatable activities (for a service to be effectively automated, it is necessary that the stakeholders agree to produce or receive it in the new fashion) or to the implications that automation may have for coordination within organizations in terms of communication between different units, resources distribution, assumption of responsibilities and reframing of professional profiles. These are factors that call into play behavioural outcomes and expectations influenced by social and psychological factors such as trust and identity, and by divergent incentives and disincentives that are difficult to reshape in a new equilibrium.

Medical imaging: an ideal target for automation?

In this view, medical imaging can be considered a 'textbook example' of a field highly compatible with automation processes. The combination of new-generation interconnected diagnostic devices (connectivity), the collection of large volumes of examinations and verified reports (Big Data) and their use for the development of machine-learning systems (AI) constitutes an ideal framework for automation processes to take place. These technologies can enhance treatments and improve performance, two major goals of both public and insurance-type healthcare systems driven by cost and profit constraints (Ratia et al, 2018); whereby an intense scientific debate has emerged in recent years, particularly in clinical and technical publications (Pesapane et al, 2018a).

Medical imaging corresponds to some families of radiological examinations based on different detection technologies: X-ray, ultrasound examinations, magnetic resonance imaging (MRI) and computed tomography (CT) (Pesapane et al, 2018b).

According to Morra et al (2019), the application of IT technologies in medical imaging has gone through three fundamental development phases:

- from the 1970s to 1990s, when pioneering 'rule-based systems' emerged;
- from the 1990s to the mid-2010s, when the use of so-called 'Computer Aided Diagnosis' became frequent;
- the current phase, characterized by the spread of 'artificial intelligence' systems.

Rule-based systems consist of applications that are able to perform rather simple procedures on the basis of indications, characteristics and typification (the 'rules') established by the user: 'if you see a white spot in the X-ray image of a certain part of the body, then report the image to the user'. Computer Aided Diagnosis (CAD) represents an evolution of the previous solutions, capable of performing more sophisticated tasks: detecting and measuring the characteristics (shape, contour, density, size and so on) of an abnormality, highlighting other relevant clinical information and supporting medical staff in identifying correlations that may help to formulate the diagnosis. CAD systems do not differ substantially from rule-based systems, since they too perform processes starting from a rationale predetermined and conditioned by human perceptual and cognitive capacities. In other words, these systems can support staff and make the reporting process more efficient, without being able to replace it completely.

The case of so-called AI, a popular term identifying a field of computer science dedicated to the creation of systems capable of mimicking human cognitive functions, such as learning and problem solving (Russell and Bohannon, 2015), is different. In the AI field, machine learning (ML) encompasses all those approaches that allow computers to learn from data (for instance, exposure to examples), such as Artificial Neural Networks (ANNs) – computational models that mimic the architecture of biological neural networks in the brain – and Deep ANNs, also called Deep Learning (DL) systems, which have emerged as promising tools. DL can in fact automatically discern relevant features even from unlabelled data, finding meaningful patterns and complex relationships often through mechanisms foreign to human logic. In medical imaging, this processing capability, coupled with the increased quantity and quality of available images, allows the detection of abnormalities usually ignored by traditional techniques and the identification of novel correlations with evidence both intrinsic and extrinsic to the specific test (for example, other clinical data).

Considering the three basic phases of performing radiologic examinations:

(1) image acquisition and processing (reconstruction, enhancement and so on),
(2) image analysis (segmentation, measurement and so on),
(3) image interpretation and drafting of the report,

the technological developments described earlier have progressively widened the area of potentially robotizable activities. However, if the application of rule-based and CAD systems concerns just the first two phases of acquisition and analysis, the most recent ML/DL systems have also reached the third phase of interpretation, allowing them to replicate the entire diagnostic process.

In fact, the automation feasibility of an entire diagnostic process based on radiological images was certified for the first time on 13 February 2018, when the US Food and Drug Administration (FDA) approved for marketing an AI software called ContaCT. This software analyses CT scans for cerebral vessel blockages while the first-line provider is conducting a standard review of the images and automatically notifies a neurovascular specialist, which potentially brings in the specialist sooner than would be the case under the ordinary standard of care. In a clinical trial, where the algorithm was tested against two neuroradiologists, the overall accuracy was 78%. But the most relevant result was the reduction of processing time: automatic notifications saved an average of 52 minutes in more than 95% of cases and the median time to notification was under 6 minutes, a time saver that can help reduce the loss of brain tissue and, ultimately, save lives. On 15 February 2018, two days after the ContaCT validation, the FDA announced the approval of an oncology software platform named Arterys to evaluate the liver nodules on MR (magnetic resonance) images and CT scans, as well as nodules on lung CT. This DL algorithm segments and measures lung and liver lesions and track tumours over time (Bluemke, 2018).

From a technical feasibility standpoint – and now also from a regulatory point of view – the complete automation of some clinical radiology processes seems to be becoming a reality. We ask: will these factors enable to a large extent the spread of radiological solutions and will they be effectively recognized by medical staff and accepted by patients? Past experiences based on previous-generation technologies, starting from the partial diffusion of CAD systems (Gurgitano et al, 2021), suggest that other aspects of a non-techno-economic nature should be considered, aspects that can be traced back to the human and relational dimension of therapeutic practices and care relationships.

AI acceptance in context: hindering factors

As shown by Wolff et al (2020), an assessment of reviewed publications on AI shows that while there is some work in this area, they have several methodological deficits. Available studies mainly focus on possible economic, organizational and social impacts, including phenomena such as: power shifts; reassignment of decision-making responsibility; cost reduction; personnel shifts and downsizing as jobs are done by robots (Mpu and Adu, 2019). Our work, however, does not deal with these impacts, but instead with the discouraging factors that might prevent the implementation of AI and digital technologies in clinical radiology. The use of AI and digital technologies will, we argue, be driven not only by technical feasibility and foreseen consequences, but also by what is considered humanly and socially desirable. Consistently, even if technical feasibility supported by AI applications increases continuously, its implementation in clinical radiology practice remains rather complex and has so far been slow (Dreyer et al, 2017). Considering earlier experiences based on CAD systems, it can be assumed that AI applications will also encounter barriers to implementation at several levels: from stakeholders and organizational procedures with strong routines and professional identities, to regulatory standards (Greenhalgh et al, 2017).

Existing theories on the uptake of technology and scientific evidence, such as technology acceptance theories (Davis, 1989), have placed emphasis on users' expectations and characteristics, and have suggested that the key determinants of technology adoption are the *ease of use* and *perceived usefulness* of the technology itself, with *social influence* added as a third determinant (Venkatesh et al, 2000). More extensively, the role of context, including cultural and professional norms, organizational infrastructures and broader social and political influences (for example, policies, laws and regulations) should be examined, assuming that acceptance by users is influenced not by isolated individuals but by the way individuals look at actions and judgements in context (Esmaeilzadeh, 2015). For example, concepts from organizational and practice theories have rarely been mobilized to explain the uptake of CAD systems. This most likely happened because CAD initially developed as an 'aid', and not as a primary decision maker. For this reason, radiologists should not dismiss their own findings if there is no CAD mark at the suspected lesion that they have found in their first reading. As a tool, CAD has never been expected to compromise the clinicians' critical reasoning, medical judgements and professional autonomy. On the other hand, CAD 'technological intentionality' (Ihde, 1990), which is entirely relational, has been neglected as an analytical approach to posit what occurs in a process of mutual constitution within which norms, values and beliefs come into play. In line with Orlikowski and Scott (2008), who posit that 'sociomaterial assemblages' entail the logical structure of 'relationality', it can be argued that the way CAD can interfere with clinicians' course of action (Liberati, 2017) recalls this important stream of research on technology and organizations.

Moreover, technologies that have been created to augment treatment for those already engaged in healthcare services such as telehealth (which provides healthcare remotely by means of telecommunications technology) are becoming increasingly common, yet the implementation and integration of these enabling technologies still need to be addressed and understood. They will presumably contribute to redefining the context of doctor–patient relationships, blurring the distinction between hospital and home care, posing additional challenges to the definition of healthcare contexts and amplifying barriers (such as issues of data storage). Therefore, health systems need to address how to best translate the enthusiasm regarding the potential of AI in everyday clinical practice (Panch et al, 2019). Talal et al (2020) found that the sociotechnical system model, with emphasis on patient-centred factors, can provide a useful interpretative framework for telemedicine deployment, implementation and barriers. According to the authors, targeting telemedicine to a vulnerable population requires the additional consideration of trust, within a comprehensive, patient-centred telemedicine system. Moreover, from an STS perspective, it is possible to think differently about how a telemedicine clinician could be observed moving from communal to intimate space (Van Tiem et al, 2021).

Tackling possible barriers calls for systematically addressing key challenges. The literature on technology implementation has recently addressed not just 'adoption' but also *non-adoption* and *abandonment* of technologies as a necessary tool to frame the uptake of computerized decision-support systems in hospitals, especially within the field of radiology. The efforts in this direction have been undertaken by several research groups (Greenhalg et al, 2017; Liberati et al, 2017; Strohm et al, 2020) who found evidence for several factors that could hinder attempts to scale up the spread of technological innovations in health and social care.

In this regard, NASSS framework (Greenhalgh et al, 2017) has been developed to produce an evidence-based, theory-informed and pragmatic framework to help predict and evaluate the success of technology-supported health and social care programmes.

NASSS was also developed to study unfolding technology programmes in real time, and in particular to identify and manage their emergent uncertainties and interdependencies.

Among its potential uses is that it can inform the design of a new technology. Previous implementation frameworks were designed around a rigid (and apparently systematic) model, but here the authors have argued that such an approach is counterproductive because eHealth technologies are typically introduced into a complex system and thus require a flexible approach. This approach has been applied to the field of clinical radiology as well, and adapted for a specific case study (Strohm, 2020), with the goal of contributing to the existing empirical evidence on the implementation challenges of AI-based medical technologies.

The domains considered in the NASSS framework, which is one of the most comprehensive efforts to face the challenges of AI adoption within medical care, are: (1) the type of medical condition involved; (2) the technology deployed; (3) the value proposition; (4) the adopter system; (5) the organization; (6) the wider context; (7) adaptation over time (Strohm, 2020). Each domain contributes to systematically grasping and orienting a possible set of determinants towards a practical evaluation of a foreseeable health or social programme and taking a dynamic perspective by following the interactions between these domains over time. Patient-centred care, within this framework, is critical, since patients, lay caregivers and staff are considered in relation to factors such as changes in staff roles, practices, identities and expectations in the implementation of technology during the diagnosis and treatment process. The staff, the patients and carers who will be expected to use AI technology as part of the suggested protocol may refuse to use it or find they are unable to use it. As an example, complexity may arise when the roles and practices assumed by the technology threaten professional traditions and codes of conduct among staff, patients and caregivers: as AI technologies increasingly take on tasks and decisions traditionally performed by humans, how should professionals consider giving up part of their agency to AI systems?

This conceptualization was the basis of subsequent studies (Strohm, 2020), capable of moving forward and identifying empirically the *lack of acceptance* of technology by staff, patients or lay caregivers, within a protocol, as one of the most important causes for non-adoption or abandonment, and thus a principal barrier to the successful implementation of AI applications in radiology.

In the wake of these studies, our contribution aims to expand further the patients' and the clinicians' perspective of the factors hindering AI and digital technology integration within radiology, taking as the fundamental basis of our epistemology a core insight expressed by Serlin (2011, p 428), who observed how 'in a world of increasingly impersonal technology, disconnection, and disembodiment a psychology that stresses connection and embodied experience is needed more than ever', in our view, factors such as trust and acceptance require to be better understood and framed as socio-psychological phenomena and to draw on a broader theoretical perspective capable of putting group dynamics, social interaction and the way information is built up at the core of the reflections concerning adoption or non-adoption of AI.

The interviews

Expert interviews were carried out, using a set of open-ended questions addressed to four radiologists from different theoretical and professional backgrounds – a project/team manager from a noted public hospital, a clinical expert (medical imaging in trauma, vascular and musculoskeletal pathology), a university professor in charge of the 'organization of the radiological service' and a clinician with a strong advocacy experience. We combined the literature review with a 'responsive interviewing' technique in order to improve our understanding of the topic and support the literature review. This method (Revell, 2013) offers insights into participants' experiences in their own words. Our open-ended questions were given to the participants before each planned interview in order to allow them to collect meaningful thoughts and ideas. At the time of the interviews, priority was given to establishing trust in such 'conversational partnerships' (Revell, 2013) for meaning-making to occur.

The questionnaire given before the interviews included a set of standard questions applied to the two main development phases of medical imaging,

CAD systems and AI systems, to bring out similarities and differences in terms of approach and implications. In particular, the following were investigated: types of examinations mostly affected by the technologies deployed, the methods of implementation in relation to pre-existing processes, job profiles and skills most affected by this implementation, changes induced in terms of productivity and cost and changes in the clinical approach, and occupational implications (for example, increase or reduction of staff). Particular attention was paid to socio-organizational relationships between staff directly involved in the diagnostic activity, between these people and other health personnel and to the relationship with patients.

After completing the interviews, we found that each interviewee provided a different path in addressing the various dimensions of the conceptual framework under analysis, which was never fully mentioned but which worked for us as a reference framework to enquire to what extent the topics raised tap into the dimensions of the framework, enriching it or possibly disregarding some of the components. In what follows we present a discussion of the dimensions that have been further enriched, and we suggest how they might be related to technology acceptance/rejection. Such an expanded framework could fill some gaps in the existing literature on technology acceptance/rejection and provide recommendations for future empirical research, highlighting directions that may help to overcome barriers in the future.

Patients' and clinicians' perspectives within an adoption system

By conceptually expanding the domain of the 'adopter system' we argue that a crucial role is determined by the *patient's perspective* as a necessary, but undervalued, component of the system, and that this perspective should be included in decisions about the design of implementation processes for AI application. The patient-centric design focuses on patients and their ideals and requires an active partnership between patients, families and staff that ensures optimal outcomes for the patient throughout their journey. Such projects do not begin with technical questions and gradually acquire social, political and economic dimensions as they progress (Callon, 1987). Rather, they are fundamentally heterogeneous and complex from the very start. This rules out the clear separation of technical and social issues in the study of technical designs (Schubert and Kolb, 2020). The symmetry proposed by actor network theory (ANT) as a solution to this separation has allowed Callon (1987) to trace heterogeneous associations and to study the socio-materiality of 'society in the making'. A second symmetrical aspect is that design should not prioritize human or technical agency alone. According to Berg (1998), there is a tendency to lean into human-centric approaches and, in so doing, to neglect the transformative role of technology in social practices (Garrety and Badham, 2004). Prioritizing the human side of design while associating technology with negative concepts of authority and control tends to result in technological designs that are configured for the minimally invasive support of existing work practices, while also preventing disruptive, albeit potentially positive, technological change (Berg, 1998).

To return to our NAASS reference framework, the patient's perspective implies a revision with respect to the determinants previously highlighted and identified in Strohm (2020) with (a) *insufficient knowledge* and (b) *trust* as determining factors of acceptance.

The full development of a 'patient perspective' adds further dimensions to the role of these determinants for AI acceptance, which will require further expansion of these dimensions to better grasp the feasibility of a foreseeable health or social programme. The articulation of emergent insights to a framework revision, well addressed by our interviewees, will likewise enhance the human dimension, with regard to the following dimensions:

(1) *knowledge* should not be considered in relation to clinicians only, but also in relation to whether AI should be explained to patients and therefore what knowledge they rely on with respect to the decision taken by the algorithm;

(2) *trust* should be considered not only as an interpersonal phenomenon but as related to intergroup distinctiveness (that is, the amount of trust placed in AI compared to human beings in accomplishing specific tasks); *trust* also applies to the clinician's perspective and will be discussed with respect to both points of view. Additional aspects, including the 'clinicians' perspective', could be further expanded on and taken up;

(3) changes in clinicians' *professional identity*. AI applications in radiology are structured to automate certain medical decision processes, thereby introducing the possibility of job displacement, which, in the field of health digitization, entails a change in the core duties of clinicians and therefore in their professional identity.

Patient's perspective

Knowledge at stake: AI explainability and the black box environment

The ethical duty of data transparency, the assumptions involved in understanding the value of the foundational data and the process by which the algorithm interprets these data for the clinical processes, together constitute the basis for offering explanations to patients. Explanations matter for patients to have confidence in their care, and for clinicians to behave in an ethical manner while providing that care. This represents a determining factor within an adoption system view. The majority of the person–centred approach entails informed decisions as a way to successfully manage patient health and care through shared decision making. In this collaborative process, through which a healthcare professional supports a patient to reach a decision about their treatment, issues of (a) *interpretability* (the ability to understand what criteria are employed to build the output), (b) *explainability* (the ability to explain how the output leads to a decision), and (c) *transparency* (the ability for a third party to understand how a decision has been taken) are expected to be tackled.

But a question arises: how is it possible to maintain *interpretability*, *explainability* and *transparency* if a lot of what happens in AI occurs in a 'black box' environment?

As to the *interpretability*, issues of race and gender must be profoundly questioned. As more organizations and industries adopt digital tools to identify risk and allocate resources, the automation of racial discrimination is a growing concern (Benjamin, 2019). Coded inequity is perpetuated if those who design and adopt AI tools are not thinking carefully about systemic racism.

As for *explainability*, according to Ghassemi et al (2021) the explanation methods developed so far are not very robust, especially when they are used to understand the functioning of the algorithm in a single case. Therefore, the black-box nature of current AI systems has caused some to question whether AI must be explainable in order to be used in high-stakes scenarios such as medicine. Current explainability methods are unlikely to achieve the levels required for patient-level decision support (Ghassemi et al, 2021). The authors provided an overview of current explainability techniques and highlighted how various failure cases can cause problems for patient decision making. In the absence of suitable explainability methods, they advocate for the rigorous internal and external validation of AI systems. Validation systems of AI-based solutions entrusted to 'third parties' also emerged during the interviews as a possible solution to ensure, if not explainability, at least accountability.

As for *transparency*, sense-making is not just about opening the closed box of AI, but also about who is around the box, and the sociotechnical factors that govern the use of the AI system and the decision. Thus the 'ability' in explainability does not lie exclusively in the guts of the AI system, but in the incorporation of the socio-organizational context (Eshan et al, 2020).

Building on the aforementioned gaps, and, as emerged from our interviews, having explainability as a requirement for clinically deployed models, means that decisions must be taken cautiously and assessed on a case-by-case basis. Thus, *personalized communication* could represent a key determinant within an 'adoption system'. It relates to the challenges in moving to the personalization of treatments based on individual patient characteristics, including molecular and behavioural biomarkers, rather than on population averages. While in the last 10 years there has been a lot of enthusiasm about the potential of 'Big Data' and ML-based solutions, the limited number of examples that impact on current clinical practice can be attributed in part to insufficient performance of predictive models and lack of validation. The limited impact of personalized medicine is also attributable to another barrier, namely, the lack of patient empowerment (Pritchard et al, 2017). In addition to 'personalized communication', a sociotechnically informed perspective that incorporates the socio-organizational context into explaining AI-mediated decision making should be conceptually and practically embedded into explainability. As addressed by Ehsan (2020), the epistemic blindspot around AI explainability could explore social transparency (ST) – the incorporation of socio-organizational context – to enable the explainability of AI-mediated decision making.

Patients' and clinicians' perspective

Trust bestowed to AI within an adoption system: dependence on task nature and the salience of human identity

It has been argued by some that explainable AI will engender trust within the healthcare workforce (Cutillo, 2020; add another reference). However, our respondents declared that trust extended towards other human beings is not likely to be automatically extended towards algorithms. The success of integrating AI into organizations critically depends on workers' trust in AI technology (Glikson and Woolley, 2020). Trust is an important factor in human–technology interaction, and the nature of tasks involved can be crucial in determining the trust workers place in a technology. In technical tasks that require data analysis, trust in AI has been found to be higher than in humans. In tasks that require social intelligence, trust in humans is higher than in AI. Moreover, drawing on social psychological research on intergroup distinctiveness, concerns about AI and digital technologies might be related to how we define and defend our human identity.

A sizable body of work from the psychology of intergroup relations has shown that both realistic and identity threats fuel intergroup prejudice (that is, negative attitudes), discrimination and conflict. Realistic threats have been shown to be an underlying factor in intergroup prejudice, discrimination and conflict (Stephan et al, 1999; Riek et al, 2006). As such, when a group's distinctiveness is threatened, this may also lead to prejudice and discrimination (Jetten et al, 2004; Yogeeswaran et al, 2012). In recent studies, authors (Ferrari et al, 2016; Złotowski et al, 2017) have examined whether exposure to autonomous robots promotes more negative attitudes towards them. Because of increased realistic threats and/or identity threats, people feel that robots pose a threat to human safety, resources, well-being and distinctiveness. Participants who watched videos of supposedly autonomous robots perceived robots in general to be significantly more threatening to humans (both realistic and identity threats), than those who watched a video of non-autonomous robots, demonstrating that autonomy affects their social acceptance. Such findings have practical implications for research on the social acceptance of AI systems as well the need for additional empirical investigations on the social perception of autonomous AI systems.

New questions about the relationship between AI/robot autonomy and their social acceptance will continue to arise in the healthcare environment. In fact, during the diagnostic process, clinicians bring into play both a capacity for logic, which is expressed through an accurate analysis of symptoms, and a capacity for social relationality, which is based on internal signals acquired with experience.

Clinician's perspective
Approaching change in radiologists' identity

Across the healthcare field, radiologists' self-identify as frontrunners for utilizing digitized supporting instruments in their everyday practice has already made them more likely to accept the redefinition of their core duties. The risk of AI and digital technologies in replacing radiologists, and thereby threatening their professional identity, has been discussed in many recent publications (Chockley and Emanuel, 2018). However, among our interviewees the phenomenon was frequently framed as more of a challenge than a threat. The option for a generational change and the possibility of developing an interventional radiology practice is part of the motivation for this view.

On the one hand, radiologists, in order to take on a leading role in the implementation of AI within the hospital, know that the next era of radiology will extend far beyond imaging interpretation and report generation. Instead, it will require a wider set of skills which could extend from data science to IT, from health systems science to enhanced diagnostics. As noted in an interview, the density and articulation of the knowledge required by the new paradigm could lead to a greater plurality of tasks and skills in the hands of radiologists, but also a clearer distinction between them and the technical staff in charge of the management of hardware and the collection of images. Considering the fact that in healthcare the supply of services contributes to determining the intensity of demand, fears of a widespread replacement of human labour did not appear to be common among the interviewees.

On the other hand, to the extent that AI literacy will be a fundamental part of the young radiologist's training, it might raise concerns for possible power conflicts with older generations that are not enabled or willing to be effectively upgraded. Knowledge about AI is advancing rapidly, whereas the duties that radiologists already have to fulfil, due to their daily commitments, do not allow many to be at the forefront of ongoing developments. This could generate a power conflict between young radiologists who hold an immaterial power, namely the knowledge, and experienced radiologists, who hold a formal type of power and status. Finding a balance between these instances of power is what will be required for the correct implementation of AI technologies to occur.

Conclusion

Our chapter has focused on the factors that prevent the social acceptance of AI technologies in the clinical setting, in particular the radiological one, which is assumed here as 'textbook example' of a field highly compatible with automation processes. The initial assumption lies in the idea that, together with impact on the quantity of employment and the quality of work practices, the factors that can prevent the successful implementation of digital technologies, especially those based on AI, must be carefully examined. The requirement outlined by the European Union to take into account the well-being of society as a whole, particularly in relation to its latest-generation policies, puts a focus on social impact, which includes possible changes in social relationships and the loss of social skills. The guidelines state that such effects must 'be carefully monitored and considered' and that AI interacting with humans must clearly signal that its social interaction is simulated.

Far from the idea that associating negative concepts of authority and control with technology can actually warn against the adverse effects of technological changes, which could also lead to positive effects, the beginning of this chapter recalled the need for an approach that shuns both reductionist positions and those that place human agency exclusively in the foreground. Rather, the chapter has aimed to articulate the debate, avoiding the sterile separation between technical and social issues, taking up the symmetry proposed by ANT as a possible solution, which in the past has allowed scholars of the STS disciplines, such as Callon (1987), to trace heterogeneous associations and to study the socio-materiality of the 'society in the making'.

Beyond purely epistemological positions, our chapter, in practice, has proposed an analysis of factors that could discourage or influence the adoption of automation in medical imaging, paying particular attention to the relationships between doctors and between doctors and patients. Compared to evidence-based frameworks such as NASSS – which is theoretically informed, sufficiently pragmatic and can help to predict the success rate of a health or social programme supported by technology – a reinterpretation of it is proposed here which deepens its scope. We have tried to understand what changes in the roles of adopters' practices, identities and expectations in implementing the technology might result in preventing the adoption of technologies. It has been found that complexity can arise when the roles and practices assumed by technology threaten established professional practices and codes of conduct among patients and healthcare professionals. As AI technologies increasingly take on more tasks and decisions traditionally performed by humans, how is human agency reconfigured within AI systems? What risks and fears should innovative diagnosis and treatment systems respond to?

The core of our analysis has focused on the conceptual, and then practical, expansion of the system of adopters: a larger picture can fill some gaps in the existing literature on technology acceptance/rejection factors and provide recommendations for future empirical research. First, taking up the factors analysed in Strohm (2020), the full development of a 'patient perspective' adds further dimensions to the role of these determinants for AI acceptance which fit within the (1) knowledge- and (2) trust-related factors. Knowledge should be considered in relation not only to what clinicians understand about AI, but also to how and if AI should be explained to patients. Trust should not only be considered as an interpersonal phenomenon but related to intergroup distinctiveness, that is, the amount of trust that the human being is willing to place in AI with respect to the performance of specific tasks that involve social interaction.

The practical implications are the following: how to communicate the algorithm's activity to patients. And in particular, how to implement the possible solutions that have been identified, namely:

- personalized communication;
- the incorporation of the socio–organizational context in order to allow for the explainability of decision–making processes mediated by AI.

Another practical implication is to examine the nature of trust in the relationship between the actors involved in AI and changes to professional identity. Both of these aspects are not sufficiently interpolated in health technology assessment. Our analysis shows that trust might be linked to a threat to human distinctiveness, and professional change to a series of beliefs about the evolution of the socioeconomic context and therefore to the way in which literature and society are dealing with the technological imagery of the future. This chapter has raised more questions than answers. Implementation of AI systems is only at the beginning, but they are already leaving a mark on government by improving the provision of essential social goods and services, from healthcare, education and transportation to food supply, energy and environmental management. The prospect that progress in AI will help society to confront some of its most urgent challenges is sometimes legitimately concerning. A steep learning curve means mistakes and miscalculations, but dealing with hindering factors and anticipating harmful impacts will make opportunities certainly more desirable.

References

Arntz, M., Gregory, T. and Zierahn, U. (2016) 'The risk of automation for jobs in OECD countries: a comparative analysis', OECD Social, Employment and Migration Working Papers, no 189, Paris: OECD Publishing.

Arntz, M., Gregory, T. and Zierahn, U. (2017) 'Revisiting the risk of automation', Economics Letters, 159: 157–160, https://doi.org/10.1016/j.econlet.2017.07.001.

Autor, D.H. (2015) 'Why are there still so many jobs? The history and future of workplace automation', The Journal of Economic Perspectives, 29(3): 3–30.

Benjamin, R. (2019) 'Assessing risk, automating racism', Science, 366(6464): 421–422.

Berg, M. (1998) 'The politics of technology: on bringing social theory into technological design', Science, Technology, & Human Values, 23(4): 456–490.

Bluemke, D. (2018) 'Radiology in 2018: are you working with AI or being replaced by AI?' Radiology, 287: 365–366, https://doi.org/10.1148/radiol.2018 184007.

Callon, M. (1987) 'Society in the making: the study of technology as a tool for the social construction of technological systems', New Directions in the Sociology and History of Technology, 83–103.

Chockley, K. and Emanuel, E. (2016) 'The end of radiology? Three threats to the future practice of radiology', Journal of the American College of Radiology, 13(12): 1415–1420.

Cutillo, C.M., Sharma, K.R., Foschini, L., Kundu, S., Mackintosh, M. and Mandl, K.D. (2020) 'Machine intelligence in healthcare – perspectives on trustworthiness, explainability, usability, and transparency', npj Digital Medicine, 3(1): 1–5.

Davis, F.D. (1989) 'Perceived usefulness, perceived ease of use, and user acceptance of information technology', MIS Quarterly, 319–340.

Dreyer, K.J. and Geis, J.R. (2017) 'When machines think: radiology's next frontier', Radiology, 285(3): 713–718, https://doi.org/10.1148/radiol.2017171183.

Ehsan, U. and Riedl, M.O. (2020) 'Human-centered explainable AI: towards a reflective sociotechnical approach', in International Conference on Human-Computer Interaction, Cham: Springer, pp 449–466.

Esmaeilzadeh, P., Sambasivan, M., Kumar, N. and Nezakati, H. (2015) 'Adoption of clinical decision support systems in a developing country: antecedents and outcomes of physician's threat to perceived professional autonomy', International Journal of Medical Informatics, 84(8): 548–560.

Eurofound (2019) Technology Scenario: Employment Implications of Radical Automation, Luxembourg: Publications Office of the European Union.

Ferrari, F., Paladino, M.P. and Jetten, J. (2016) 'Blurring human–machine distinctions: anthropomorphic appearance in social robots as a threat to human distinctiveness', International Journal of Social Robotics, 8(2): 287–302.

Frey, C. and Osborne, M. (2013) *The Future of Employment: How Susceptible Are Jobs to Computerization?* Oxford: Oxford Martin School.

Garrety, K., Robertson, P.L. and Badham, R. (2004) 'Integrating communities of practice in technology development projects', *International Journal of Project Management*, 22(5): 351–358.

Ghassemi, M., Oakden-Rayner, L. and Beam, A.L. (2021) 'The false hope of current approaches to explainable artificial intelligence in health care', *The Lancet Digital Health*, 3(11): e745–e750.

Glikson, E. and Woolley, A.W. (2020) 'Human trust in artificial intelligence: review of empirical research', *Academy of Management Annals*, 14(2): 627–660.

Greenhalgh, T., Swinglehurst, D. and Stones, R. (2014) 'Rethinking resistance to "big IT": a sociological study of why and when healthcare staff do not use nationally mandated information and communication technologies', Southampton: NIHR Journals Library.

Greenhalgh, T., Wherton, J., Papoutsi, C., Lynch, J., Hughes, G., Hinder, S. et al (2017) 'Beyond adoption: a new framework for theorizing and evaluating nonadoption, abandonment, and challenges to the scale-up, spread, and sustainability of health and care technologies', *Journal of Medical Internet Research*, 19(11): e367.

Gurgitano, M., Angileri, S., Rodà, G., Liguori, A., Pandolfi, M., Ierardi, A. et al (2021) 'Interventional radiology ex-machina: impact of artificial intelligence on practice', *La radiologia medica,* 126, 10.1007/s11547-021-01351-x.

Hsu, H. and Pinch, T. (2008) 'Affordances and theories of materiality in STS', www.academia.edu/34067425/Affordances_and_Theories_of_Materiality_in_STS.

Jetten, J., Spears, R. and Postmes, T. (2004) 'Intergroup distinctiveness and differentiation: a meta-analytic integration', *Journal of Personality and Social Psychology*, 86(6): 862.

Keynes, J.M. (2010) 'Economic possibilities for our grandchildren', in *Essays in Persuasion*, London: Palgrave Macmillan.

Ihde, D. (1990) *Technology and the Lifeworld: From Garden to Earth*, Bloomington, IN: Indiana University Press.

Liberati, E.G., Ruggiero, F., Galuppo, L., Gorli, M., González-Lorenzo, M., Maraldi, M. et al (2017) 'What hinders the uptake of computerized decision support systems in hospitals? A qualitative study and framework for implementation', *Implementation Science*, 12(1): 1–13.

Morra, L., Delsanto, S. and Correale, L. (2019) *Artificial Intelligence in Medical Imaging: From Theory to Clinical Practice*, Boca Raton, FL: CRC Press.

Mpu, Y. and Adu, E.O. (2019) 'Organizational and social impact of artificial intelligence', *American Journal of Humanities and Social Sciences Research*, 3(7): 89–95.

Nedelkoska, L. and Quintini, G. (2018) 'Automation, skills use and training', OECD Social, Employment and Migration Working Papers, no 202, Paris: OECD Publishing.

Nikishawa, R.M. and Bae, K.T. (2018) 'Artificial intelligence may cause a significant disruption to the radiology workforce', *Journal of the American College of Radiology*, 16(8):1077–1082.

Orlikowski, W.J. and Scott, S.V. (2008) 'Sociomateriality: challenging the separation of technology, work and organization', *Academy of Management Annals*, 2(1): 433–474.

Panch, T., Mattie, H. and Celi, L.A. (2019) 'The "inconvenient truth" about AI in healthcare', *npj Digital Medicine*, 2(1): 1–3.

Pesapane, F., Codari, M. and Sardanelli, F. (2018a) 'Artificial intelligence in medical imaging: threat or opportunity? Radiologists again at the forefront of innovation in medicine', *European Radiology Experimental*, 2(1): 1–10.

Pesapane, F., Volonté, C., Codari, M. and Sardanelli, F. (2018b) 'Artificial intelligence as a medical device in radiology: ethical and regulatory issues in Europe and the United States', *Insights into Imaging*, 9, https://doi.org/10.1007/s13244-018-0645-y.

Pritchard, D.E., Moeckel, F., Villa, M.S., Housman, L.T., McCarty, C.A. and McLeod, H.L. (2017) 'Strategies for integrating personalized medicine into healthcare practice', *Personalized Medicine*, 14(2): 141–152.

Ratia, M., Myllärniemi, J. and Helander, N. (2018) 'Robotic process automation-creating value by digitalizing work in the private healthcare', *Proceedings of the 22nd International Academic Mindtrek Conference*, pp 222–227.

Revell, S.M.H. (2013) 'Making meaning in qualitative research with conversational partnerships: a methodological discussion', *Advances in Nursing Science,* 36(2): E54–E65.

Riek, B.M., Mania, E.W. and Gaertner, S.L. (2006) 'Intergroup threat and outgroup attitudes: a meta-analytic review', *Personality and Social Psychology Review*, 10(4): 336–353.

Russell, S. and Bohannon, J. (2015) 'Artificial intelligence: fears of an AI pioneer', *Science*, 349: 25.

Schubert, C. and Kolb, A. (2021) 'Designing technology, developing theory: toward a symmetrical approach', *Science, Technology, & Human Values*, 46(3): 528–554.

Serlin, I. (2011) 'The history and future of humanistic psychology', *Journal of Humanistic Psychology*, 51(4): 428–431.

Stephan, W.G., Stephan, C.W. and Gudykunst, W.B. (1999) 'Anxiety in intergroup relations: a comparison of anxiety/uncertainty management theory and integrated threat theory', *International Journal of Intercultural Relations*, 23(4): 613–628.

Strohm, L., Hehakaya, C., Ranschaert, E.R., Boon, W.P. and Moors, E.H. (2020) 'Implementation of artificial intelligence (AI) applications in radiology: hindering and facilitating factors', *European radiology,* 30: 5525–5532.

Talal, A.H., Sofikitou, E.M., Jaanimägi, U., Zeremski, M., Tobin, J.N. and Markatou, M. (2020) 'A framework for patient-centered telemedicine: application and lessons learned from vulnerable populations', *Journal of Biomedical Informatics,* 112, 103622.

Van Tiem, J.M., Reisinger, H.S., Friberg, J.E., Wilson, J.R., Fitzwater, L., Panos, R.J. et al (2021) 'The STS case study: an analysis method for longitudinal qualitative research for implementation science', *BMC Medical Research Methodology*, 21(1): 1–12.

Vernoni, G. (2018) 'La quarta rivoluzione industriale e gli effetti sul lavoro: una chiave di lettura', in C. Lupi (ed) *Il futuro della fabbrica*, Milan: Este Libri, pp 25–41.

Venkatesh, V. and Davis, F.D. (2000) 'A theoretical extension of the technology acceptance model: Four longitudinal field studies', *Management Science*, 46(2): 186–204.

Violante, G.L. (2008) 'Skill-biased technical change', *The New Palgrave Dictionary of Economics,* 2: 1–6.

Wolff, J., Pauling, J., Keck, A. and Baumbach, J. (2020) 'The economic impact of artificial intelligence in healthcare: a systematic review', *Journal of Medical Internet Research*, 22(2): e16866.

Yogeeswaran, K., Dasgupta, N. and Gomez, C. (2012) 'A new American dilemma? The effect of ethnic identification and public service on the national inclusion of ethnic minorities', *European Journal of Social Psychology*, 42(6): 691–705.

Złotowski, J., Yogeeswaran, K. and Bartneck, C. (2017) 'Can we control it? Autonomous robots threaten human identity, uniqueness, safety, and resources', *International Journal of Human-Computer Studies*, 100: 48–54.

8

Robots for Care: A Few Considerations from the Social Sciences

Miquel Domènech and Núria Vallès-Peris

Introduction

According to the scientific, technical and media literature, there is a continuous and significant increase in the narrative about the importance of robots in the economy and society (Belanche et al, 2020; Rommetveit, van Dijk and Gunnarsdóttir, 2020). Originally, robots were produced to cover the needs of industry, being regarded as a substitute for human beings and for doing the 'three Ds' of labour: dirty, dangerous and demeaning. In industry, robots are supposed to improve the quality of work by taking over tasks that are very hard or not safe for humans to perform (Royakkers and van Est, 2015; Vallès-Peris and Domènech, 2020a). Thus, most of the known industrial scenarios do not need to take into consideration how to manage the interaction between humans and robots; once the robot is placed, the human being is thought to disappear.

Nevertheless, the idea that robots can cover other human needs than those related to industry has also gained ground and new concerns and debates are taking place. Roboticists refer to this innovative landscape as 'New Robotics' (Schaal, 2007). Together with a set of technical developments, New Robotics contains a narrative aligned with the unstoppable and exponential development of robots, as well as a certain technological determinism that foresees a future of robots as the best solution to our problems (Maibaum et al, 2021; Rommetveit, van Dijk and Gunnarsdóttir, 2020). Inevitably, this narrative implies that robots will appear in environments that would have seemed unexpected not long ago, assuming duties traditionally and exclusively assigned to humans. In this vein, we are witnessing an increasing expectation of advances in robotics in different domains, among them in the field of healthcare.

The possibility of involving robots in caring tasks seems closer than ever. There have been pilot experiences with robots and older people, robots working in therapy for children with autism and robots in hospitals enrolled in a variety of tasks. The response to the emergency situation provoked by COVID-19 has increased the use of robots, intensifying the growing tendency of recent years. After public safety, the clinical care domain has the largest set of uses of robots, due to robots' potentiality to be used to reduce physical contact between people (Kritikos, 2020). For example: telepresence robots for patient–doctor interactions without contact; or disinfection, prescription- and meal-dispensing robots, which allow the performance of essential day-to-day tasks in a hospital without human contact, with machines navigating autonomously through a hospital (Cardona et al, 2020; Murphy et al, 2020). According to the International Organization for Standardization, robots a distinction can be made between industrial or service robots. Service robots are a wide category that includes different types of robots used in caring contexts. According to the last report of the International Federation of Robotics (IFR, 2021), service robots have a tremendous market potential worldwide. Sales of service robots rose 41% to 131,800 units in 2020, driven by the extra demand of the global pandemic (IFR, 2021). The five top types of service robots sold during the pandemic were those involved in caring contexts and activities: autonomous mobile robots and delivery robots, used in hospitals for delivering food or medicines; professional cleaning robots for disinfecting, spraying disinfectant fluids or using ultraviolet light; medical robotics, mainly driven by robotic surgery devices; robots for rehabilitation and non–invasive therapy; and social robots, designed to help, for example, residents of nursing homes to keep contact with friends and family members in times of social distancing; and hospitality robots, which provide information in public environments to avoid personal human contact (IFR, 2021).

Although a small number of studies have explored the views of the various agents involved in healthcare, existing studies analysing attitudes towards the use of robotics in health environments reflect mixed views. Robots are considered appropriate for different work tasks, while the criticism is directed towards decreasing levels of human contact and the unnecessary deployment of new technology (Savela et al, 2018). Healthcare workers consider an array of tasks that robots could perform, and highly approve of telepresence tasks, while they address the issue of patient safety and raise privacy concerns (Papadopoulos et al, 2018). Patients' perspectives are also ambivalent: on the one hand, they prefer to be cared for and to perform care practices with humans; on the other hand, they consider that it could be very positive to introduce robots into the care of human beings, assisting carers and medical personnel and, if necessary, replacing them (Vallès-Peris et al, 2021). In the field of gerontology, where there is a relatively long tradition of the introduction of companion robots and socially assistive robots, studies show that older people engage with social robots and adopt them in their daily lives (Šabanović and Chang, 2016; Khosla et al, 2021).

In what follows, we will try to show the reason for this interest in the possible use of robots for care and the arguments used to justify it. From this overview, we present an approach to the robotics of care from the social sciences, particularly from science and technology studies (STS). From this perspective, we propose a discussion around the assemblage of relationships that configure and are configured by care robots, as an open discussion to analyse the effects of robots on the relations and practices of care.

Why robots for care?

There is a general trend in scientific literature, political agendas, funding programmes and the media, which assumes that future societies will witness an increase in the introduction of robots in healthcare. However, how is such a need to develop robots for care explained?

The need to boost development and research in healthcare robotics is based on four types of arguments: (1) the care crisis; (2) the improvement of efficiency; (3) the new potentials robots offer; and (4) the market.

The care crisis

The alert that Western societies are heading towards a 'care crisis' has been commonplace for some years now, not only in the specialized literature but also in the press and the general media. The core idea of the so-called care crisis is based on the fact that contemporary healthcare systems face substantial challenges due to increasing life expectancy, an expansion in the need for health services, not enough people to provide these services and not enough resources to finance them. Studies from different agencies and bodies are cited which insist on data that reinforce this thesis. Two types of statistics that are often brought up and are, in fact, interconnected.

On the one hand, we have statistics relating to forecasts about the growth of the elderly population. The World Health Organization (WHO) estimates that between 2015 and 2050 the pace of population ageing will be much faster than in the past and the proportion of the world's population aged over 60 years will nearly double, from 12% to 22%, and the number of people aged 60 years and older will outnumber children younger aged than five years (WHO, 2021). On the other hand, and intrinsically related to the first set of statistics, this situation will entail an increase in care needs, and an increase in the number of people affected by neurodegenerative diseases should be expected on account of the ageing of the population. According to WHO data, more than 55 million people worldwide were living with dementia in 2021, and there are nearly 10 million new cases every year (WHO, 2021b).

The conclusion reached from these forecasts is that humanity is ageing, and there might be no one to take care of older people because, as it seems, there will be a shortage of health workers in the coming years. Roboticists are aware of this need: 'The number of people with partial impairments is very large and growing', says Conor Walsh, a robotics expert at Harvard University in an interview for *QMayor Magazine*, pointing out that about 10% of people living in the US have difficulty walking. According to this magazine report, 'Walsh and other researchers are working in labs across the world to ensure that these technologies not only exist, but that will be reliable, durable, comfortable and customised to users' (Redacción QMayor, 2016).

During the last decade, we have witnessed a flurry of news regarding different experiences related to the introduction of robots in nursing homes. Thanks to journalists, we know, for example, that older people at the Shin-tomi nursing home in Tokyo are not woken up every morning by an alarm clock, but by Sota, a little blue-and-white robot placed on their bedside tables. And not only this, besides saying good morning and reminding them of their medication, Sota is equipped with an infrared camera to detect if the residents fall out of their beds, and can warn their carers (Díaz, 2017). It is possible to find 20 different models of robots to care for Shin-tomi residents (Foster, 2018). According to the nursing home's director, Kimiya Ishikawa, this private facility, which has been in operation since 2001, 'is a model in Japan because it is at the forefront of technology to improve efficiency and reduce the workload of employees' (Díaz, 2017).

Besides the arguments developed earlier, it is important to keep in mind other considerations regarding caring for older people. The traditionally unfair labour conditions of carers, whether they are informal carers (mainly family members) or workers in the care sector (in health, social work and domestic services) are well known (Jokela, 2019). The lack of sufficient public resources to organize and perform care and caring (Fine and Tronto, 2020), in a sector with high exploitation and low salaries (especially for migrant women), has produced a heterogeneous collective of carers characterized by its precariousness (Lightman, 2022). These poor working conditions contrast with the image of progress and innovation that accompanies the development of digitalization. What would be, in this *décalage*, a more suitable solution than robots for care?

The improvement of efficiency

Alongside the growing demand for assistance (a situation intensified by the COVID-19 pandemic), there is a tendency towards increasing costs, worsening outcomes and healthcare systems' oversaturation, exemplifying a paradox of more investment of human capital and worse human health outcomes (Roser, 2017; Topol, 2019). This global scenario occurs at the same time as the production of data in massive quantities, as well as a new generation of machines and techniques to analyse these metrics and the compliance of new data. Robotics and its algorithms represent the integration of the need to support healthcare systems with 'more hands' as well as to provide more efficient mechanisms to reduce diagnostic errors, mistakes in treatment, waste of resources, inefficiencies in workflow, inequities and inadequate time for interaction between patients and clinicians. The need for more efficient and effective health systems demands complex solutions, including a progressive digital transformation of healthcare systems. The development of robotics is integrated into this process, with a multiplicity of robots with different functions for healthcare settings such as hospitals or nursing homes being developed.

This is the case with therapeutic robots, which are designed to motivate and assist patients to build a positive attitude towards fighting a disease (Larriba et al, 2016). An example of a therapeutic robot is the Pleo robot, a baby dinosaur robotic pet, which works in different ways to assist children during hospitalization at a paediatric hospital in Barcelona. Pleo belongs to a group of robots used to reduce pain and anxiety in hospitalized children. It has the appearance of a pet; thus, its effects can be categorized as similar to those of animal-assisted therapy, with the difference that animals are not always available, concerns about allergies or immune-suppressive conditions and the need for trained professionals, which restricts their use. Research in paediatric hospitals has concluded that using therapeutic robots with patients achieves results that are as positive as those of zootherapy, for example in the treatment of pain and anxiety or attending to the social needs of hospitalized children (Díaz-Boladeras et al, 2016) Furthermore, it has been proven that these robots not only benefit and improve the quality of life of the children, but also provide a solid support for the parents and the medical team, especially during the most delicate moments of the treatment process (González-González, Violant-Holz and Gil-Iranzo, 2021).

In a similar vein, inspired by research into the therapeutic effects of human–animal interaction, we can also find several examples of using robotic pets to support older people with dementia, with PARO, probably, being the most-used assistive robot of this kind. In a study with hospitalized patients with dementia using PARO to support their care experiences, it was concluded that patient participants perceived the robot PARO as helpful in supporting their psychosocial needs for inclusion, identity, attachment, occupation and comfort (Hung et al, 2021). According to these authors, the robot would make the hospital environment more supportive for older people.

Together with this type of therapeutic robot, some robots can be generically grouped in the category of 'nursing robots' or 'robot nurses', robots designed to help perform tasks traditionally carried out by nurses or auxiliary health personnel. The development of these robots is understood as offering solutions to the lack of care workers, as well as improving their working conditions and making the existing healthcare system more effective. In a scenario where nurses and other healthcare personnel caring for patients in overcrowded hospitals, it is said that robots can alleviate difficult situations by replacing humans in heavier or repetitive tasks. Here the introduction of robots is possible because the classic industrial imaginary that sees robots as a good solution for performing 'triple D' tasks is applied to care, which is viewed as an activity that can be broken down into simple tasks. In other words, the care robot entails a conception of care as an activity that can be fragmented into different tasks, while at the same time hierarchizing these tasks: those considered less valuable can be delegated to the robot (Vallès-Peris and Domènech, 2020a). A paradigmatic example of this type of robot is those designed to replace hoists for lifting and transferring patients. Robots are being developed that can lift/set down patients from/onto their bed or wheelchair, affording a more delicate machine–patient relationship than that provided by hoists. In this vein, assistive robotics is tending to design more closely to social robotics, which allows for the performance of such tasks in a less mechanical way (Magnenat-Thalmann and Zhang, 2014). For example, for the task of lifting and moving people with reduced mobility, the goal is to abandon hoist-like designs and to propose models that aim to be familiar, friendlier and include more features. An example of this would be the 'RIBA' robot (Ding et al, 2013), which operates by moving people in a manner analogous to that of a human being.

The new potential robots offer

Alongside robots designed to improve task performance or working conditions, another narrative accompanying the introduction of robots into care is that they offer new benefits and functionalities. Robots could offer solutions to currently unsolved problems, thus displaying their full innovative and disruptive potential. The notion of novelty is embedded in robotics, as the expression 'New Robotics' shows. New robotics embraces a broad interdisciplinary approach focused on human–centred machines designed to interact with people in daily life. As Stefen Schaal explains, 'elder care, physical therapy, child education, search and rescue, and general assistance in daily life situations are some of the examples that will benefit from the New Robotics in the near future' (Schaal, 2007, p 115). New robotics transform the imaginary of the industrial robot, designed to replace or assist humans in certain tasks. From this perspective, new machines are defined by their capacity to collaborate and interact with humans. We are no longer talking about isolated robots in industrial production plants, but about a series of developments that allow robots to participate in social life, enabling new forms of interaction to solve certain existing problems.

This is the case of robots that attend school in place of sick kids. In 2017, CNN published an interview with Leslie Morissette in which she explained how the foundation created by her, Grahamtastic Connection, had helped children hospitalized with cancer to stay connected to the outside world and to continue their education and relationships with teachers and friends. Since 1998, using tablets and laptops, the foundation has been helping to connect kids to their classrooms, and since 2012 it has been providing robotics. According to Morissette:

> the robots transport the child right into the classroom in real time. They can operate the robot right from their hospital bed or home. So, if a child is unable to attend school, they can simply log on to their tablet or laptop and call in to the robot. They can walk up and down the halls. They can go to lunch with their friends. The real magic happens between classes, when they're walking down the hallway with their friends, by robot, talking about their weekend and their favourite foods and just all the kid stuff. It connects them to their friends, their classmates, and their teachers. It's neat technology that really gives children the feeling of control, where their world is maybe out of control. (Klairmont, 2017)

Some of these experiences have already been documented in the scientific literature (Kristoffersson et al, 2013).These types of robots go beyond the idea that robotics could be useful for improving healthcare systems' efficiency or alleviating the labour conditions of care workers. This is so because New Robotics respond to an old dream of artificial intelligence to create an artificial system with similar abilities close to those of humans. However, although in public debates and academic discussions robotics are interpreted as an omnipotential solution for all problems (which can lead to either a positive or negative assessment), the current aim of researchers in New Robotics is not so ambitious. Researchers are focusing on making progress in very well-defined problem areas, such as, robotic physical therapy or robotic education of children with special needs (Schaal, 2007). For example, in the field of socially assistive robots, the focus is on the potential added value of therapy robots or educational goals for children with autism spectrum disorder (ASD). Children with ASD have particular difficulty in expressing emotions, usually have limited social skills and face major problems when communicating. In some cases, they are stressed by human behaviour, which causes them anxiety because it is unpredictable. Research centres around the world have created robots using artificial intelligence with different objectives, such as to help children with ASD to recognize facial expressions in themselves and others or foster development of social skills (Heerink et al, 2016).

There is a market

Last but not least, in the present stages, robotics is becoming an integral element of globalized economic life, a global industry that is rapidly developing. According to the International Federation of Robotics (IFR), the world market for professional service robots grew strongly in 2019 by 32%, from USD8.5 billion to USD11.2 billion, and it is expected to continue growing at rates of between 30% and 40% per annum in the coming years (Duch-Brown, Rossetti and Haarburger, 2021). Other predictions suggest that by 2030 professional services robots will reach a market volume of USD90 billion to USD170 billion and will outpace conventional industrial robots and cobots (Lässig, 2021). Although predictions differ on the volume of business, there is no doubt that the robotics market is experiencing exponential growth worldwide. It has been experiencing a significant transformation, with robotics growing beyond the workhorses of the industrial shop floors and beginning to be more relevant in the services sector, adopting the roles of personal assistants, surgical assistants, autonomous vehicles, delivery vehicles, exoskeletons and crewless aerial vehicles, among multiple other uses.

This process has been intensified by the COVID-19 pandemic, with high demand for robotic disinfection solutions, robotic logistics solutions in factories and warehouses, and robots for home delivery. Demand for professional cleaning robots grew by 92% in terms of units sold and by 51% in terms of turnover (IFR, 2021). Turnover of medical robotics was mainly driven by robotic surgery devices, which are the most expensive type of service robot. Hospitality robots enjoy growing popularity (IFR, 2021). In this context, the competition of the world's leading states for leadership in the robotics market forms a high level of relevance for research in this area of the world economy (Makedon et al, 2021). New start-up companies appear every year, developing innovative service robot applications and improving existing concepts. Forty-seven per cent of service robot suppliers are from Europe, 27% from (North) America and 25% from Asia (IFR, 2021).

Today, the economic perspective on robotics seems quite relevant; however, it fails to illuminate how, and under which circumstances, it is or will be of concern for the greater population or how it will contribute to a more equitable way of distributing resources around the world. How do this expanding robotics market and the so-called care crisis fit together? Does innovation in robotics mean a substantial transformation in the field of care? How to integrate these elements that explain the development and implementation of robotics in the field of health into an ethical reflection on robotics?

The social sciences' contribution

Given the scenario already described, the question that arises, and that will be addressed here, has to do with the role that the social sciences should play in analysing this perspective of a society that is moving towards the robotization of care. In the previous section, starting from the question 'Why care robots?', instead of focusing on the characteristics or functionalities that these artefacts may have, we have identified some demographic, economic, political and technological processes that are intertwined with the development of robotics. In this way, we understand the robot as more than a mere isolated artefact, but as a network of relationships encapsulated in an artefact, in which, at the same time, that robot participates. This approach is not original but is based on the research theory developed by science and technology studies (STS).

Robots as artefacts embedded in a network

Some of the reflections that have been elaborated in technology and technological innovations from STS can be frankly useful in addressing some of the debates that care robots raise, as well as in providing new insights for developing responsible robotics for well-being and the common good.

Understanding technology

The first contribution that social scientists can make to the analysis of the process of robotization of care is to clarify an essential question about technology, concerning its effects on social life. It is indeed almost commonplace for us to say that technology is not neutral (Miller, 2021). However, all too often, in presentations and public debates, assertions to the contrary can be heard and are often well received because, admittedly, this also seems to be the direction of common sense.

There have been social scientists who have insisted that technology is neither good nor bad; and that may seem to mean that it is neutral. Effort, therefore, must be put into clarifying the role of technology in social life and into reaffirming its lack of neutrality.

In this vein, STS have provided powerful arguments to disarticulate this technology-neutral view. Regarding the issue of care, we have a good number of studies that show how technologies of care give rise to particular effects when they are implemented (Schillmeier and Domènech, 2010). Indeed, technologies can transform relationships, affect people's identities or change meanings, among many other possibilities.

According to this STS approach to technology, the apparatuses and devices that populate our daily lives, far from being mere passive instruments that we use for different human purposes, are active agents that structure our lives. Some authors speak of 'technological culture' to refer to this agentic capacity of technology (Bijker, 2009). In speaking of 'technological culture' it is assumed that contemporary societies are fully technological and that all technologies are, in a generalized way, cultural, to emphasize that technologies are, in the end, powerful forces that reshape human activities and their meanings. Thus, 'when a sophisticated new technique or instrument is adopted in medical practice, it transforms not only what doctors do, but also the way people think about health, illness, medical care and even death' (Bijker, 2009, p 67).

Another way to refer to this role of technology in human societies is mediation (Latour, 1994). Indeed, for Latour (2005), the difference between an idea of technological devices as passive elements subject to human will and other ideas of these same devices as actors that affect the interactions in which they are immersed is the difference between considering them as intermediaries or as mediators. If an intermediary is a mere vehicle for transporting a meaning without transforming it, 'mediators transform, translate, distort, and modify the meaning or the elements they are supposed to carry' (Latour, 2005, p 39).

This approach emphasizes the relational and contextual nature of the robot, which, like any other technological artefact, becomes meaningful only when integrated into a network of relationships. That is, a robot should be understood not just as a mere artefact, but as a kind of 'black box' that encloses a whole network of devices, processes and actors. A robot can certainly be a mere assistant that goes from room to room handing out medication to patients in a hospital. However, seen from an STS perspective, the robot contains, so to speak, material elements such as cables and chips that constitute it as such, but it is also made up of protocols for action, algorithms that set guidelines for patient relations, interests of technological companies, discourses on solutions to the problems of lack of specialized personnel and so on. In short, it is a conglomerate of heterogeneous elements: material, social and semiotic.

This does not imply, for example, that distinctions cannot be drawn between a nurse and the measuring devices she uses, or between a robotic arm and a surgeon's hand, but that distinctions between elements cannot be used as principles of explanation, because what is constantly being produced are socio-techno-natural entanglements (Callon and Latour, 1991). According to considerations of this kind, defining care robots solely based on their technical characteristics, their appearance or their functionalities would not make much sense, because the robot-artefact is not the privileged actor for reflecting on the effects of robotics on our ways of life; rather, the focus of attention is on the care relations in which this artefact participates (Vallès-Peris et al, 2018; Lipp, 2022; Nickelsen et al, 2022). Van Wynsberghe (2015) takes up this idea in his definition of care robots, based not on their functionalities but on their relational component: 'care robots as artefacts designed for integration in care practices and therapeutic relationships, to meet the care needs of those who provide care or those who are its recipients, and which are used in health-care contexts such as a hospital, a residence, a day care centre or the home' (van Wynsberghe, 2015).

The effect of technological innovations

Based on this idea of technology as a mediator, a technological innovation is not simply a solution to a given problem – the care crisis, for example – but, rather, any innovative process proposes a model of a future society in line with the expected role of such innovation. Thus, a question concerning the consequences of the introduction of a technological advance is always appropriate. A society with robots is not going to be the same as a society without them. There is, then, a certain vision of society entailed by every robot offered for a particular end. This is why Callon (1987) often refers to engineers as engineer-sociologists. Through the analysis of different design processes, Callon (1987) concludes that for technological innovation to have a successful implementation, it is not only necessary that it functions and serves to solve a problem, but also that it finds a society where it has a place. It is in this sense that, to succeed, technological innovations must be designed in such a way that they contribute to making the proposed model of society a reality.

Of all the literature that addresses this question of the performative role of technologies, reference to the notion of script (Akrich, 1992) is particularly influential in the healthcare field (Nickelsen and Simonsen Abildgaard, 2022). What this approach suggests is that artefacts embody scripts, through which they influence the actions of the humans with whom they relate. These scripts are, mainly, prescriptions for use and are inscribed into the artefacts by designers and engineers.

Verbeek (2006) develops this idea of the mediating capacity of technology in conjunction with the role of scripts to argue that artefacts are materialized morality. Briefly, his argument is that designers delegate specific responsibilities to artefacts, such as the responsibility to make sure that no one drives too fast, which is delegated to a speed bump. To do that, they use scripts, which encourage specific actions (for example, slow down a car's velocity in the case of the speed bump), while discouraging others (going faster). Additionally, given that ethics, after all, is about how we act, if technologies seem to provide material answers to this issue, then we can conceive the work of designers as an inherently moral activity. Similar to Akrich's concept of script and integrating Verbeek's materialized morality, but emphasizing its political component, Feenberg develops the notion of the 'technical code' to explain how social values, ideologies and power relations are pre-inscribed in artefacts (Feenberg, 2010). However, these pre-inscriptions are unique not to the artefact, but to the artefact in an environment of production and use that is deeply biased by relations of power and domination – what Feenberg calls the instrumentalization thesis. The idea behind the instrumentalization thesis is that technologies emerge as the contingent outcome of struggles and negotiations, with the artefact and its technical code being shaped by relationships (Feenberg, 2010).

As we have already shown, robots are being designed to perform a variety of tasks in healthcare settings, such as enabling new uses and activities (for instance, facilitating patients' relationships with spaces outside the hospital or reducing their anxiety in the pre-operative period) or modifying certain actions of healthcare staff (such as measuring vital signs, performing detailed interventions and so on). Integrating the notions of mediation and instrumentalization in robotics, it could be said that when a robot is introduced into the network of care organization and practices of a hospital, the robot and the care relationships are (co-)transformed: the way that is understood and practised in care and care processes is changed, just as the materiality of the robot is reconstructed, based on the assembly with the social relationships in which the robot participates (Law and Mol, 1995).

This approach does not fit with the traditional conceptualization of human–robot relations in a binary model. The binary model is deeply rooted in the imaginary around robots, reflected in concepts in both academic and dissemination areas: such as human–robot interaction (HRI) or human–machine relations. According to this type of logic, the relationship between humans and robots is based on the idea of two independent entities that interact with each other. According to the approach we are presenting here, such a proposal does not make much sense. We prefer to conceptualize robots based on their integration in networks of care relationships, a point of view that we have identified, on other occasions, as a robot embedded in a network (REN) (Vallès-Peris and Domènech, 2021; Vallès-Peris, 2021).

The ethical debate

These arguments lead us directly to consider the need for further reflection on care robots regarding those aspects related to the ethical dimension of technology, transcending the common-sense idea that technologies should be morally evaluated only in terms of the goals for which they are designed or the quality of their performance. From a REN's perspective, the interest of ethical reflection does not focus exclusively on the moral evaluation of the robot's activities or on the analysis of its tasks and functionalities, but on the study of how robotics contributes to transforming care relations as well as how the ways of organizing, practising and valuing care are inscribed in the robot. It should be made clear that such a reflection is by no means easy, insofar as, when it is approached, considerations about developments that are real are often intermingled with purely speculative ideas. As has been pointed out, the ethics of robotics runs the risk of falling into a speculative practice that loses sight of the most current developments of the technology in its everyday context, less spectacular than the futuristic visions of certain promoters of robotics, but more affected by the pressure of the often neglected normative aspects of the here and now that really affect potential users (van der Plas, Smits and Wehrmann, 2010).

About deception and interdisciplinarity

An extensive literature is beginning to develop around the ethical issues surrounding the use of robots in general and in the field of caregiving in particular. Often, the conclusions reached refer to very general principles. Thus, for example, Lazar, Thompson, Piper and Demiris (2016), when discussing service robots intended for older people, specify that these should fit the ecology of older people's lives, enable flexible use and adapt to possible changes in functioning, while preserving older people's values, such as dignity and independence. One of the concerns that has received the most attention in the literature on the ethics of robotics has to do with the possibility that users may feel deceived by attributing capabilities or intentions to robots that they lack. This is a question of interest, since such deception would have to do with the tendency people show to anthropomorphize the robots they interact with and, as Sharkey and Sharkey (2011) point out, some robot designs are made to produce or reinforce this anthropomorphizing tendency. In some such cases, it could perhaps be argued that it is the design itself that contributes to the deception of the user, a position supported, for example, by Sparrow and Sparrow (2006), who consider that the beneficial effects of the interaction of elderly people with robotic pets comes from the deception of making them believe that the robot is something with which they can maintain a relationship. There is no doubt that, in light of what has been discussed here about the script and action mediation, it is relevant to elucidate to what extent a robot script incorporates this pattern of deception in order to achieve greater attachment to the robot.

In any case, it seems necessary to ensure that users are provided with sufficient information to be able to make informed decisions about their interactions with robots. The question to be asked in this respect, as pointed out by Feil-Seifer and Matari (2011), would be: has a careful description of the robot's capabilities been provided? And, it could be added, has the description been made comprehensible for all kinds of cognitive circumstances and ages of the recipients of the explanation?

Furthermore, the necessary information does not only refer to the patient users, but to the various actors involved in the care relationships with robots in a particular situation. For example, when a nurse is bathing a person with reduced mobility with the help of a robot, which assists her in holding and cleaning the patient, the ease of execution of the task that the robot incorporates cannot be to the detriment of the visual and bodily relationship that guarantees the good care relationship at that moment (Mol et al, 2015).

From an ethical study based on the centrality of the assemblage of care relationships in which the robot participates, we can identify which variables or criteria are relevant to facilitate 'good care' in that bathing situation. This approach not only allows us to discuss the best way to introduce these artefacts in particular scenarios and for specific tasks, but also opens the door to the discussion of care issues in a field (that of technological design) that is usually far from this type of debate and reflection. In this sense, it is ethically responsible that the processes of research and development of robotics incorporate sufficient information on care relationships to make informed designs that guarantee 'good care'. Knowledge about care from disciplines such as social sciences, humanities or nursing therefore seems essential to be integrated as an ethical requirement into the engineering and technical knowledge involved in the development of robotics.

About privacy and fragmentation

Privacy is another major concern found in the literature. This is an issue that arises when considering the ability of robots to record and store information, especially when performing monitoring tasks. It is understood that care robots throughout their interaction with users collect information about them that, in some cases, may be sensitive. What to do with this information? With whom should it be shared? Sharing patient health information with caregivers can have both positive and negative effects. On the one hand, the fact that the robot is in charge of collecting and managing certain data can be a relief for the informal caregiver, who can thus focus on other things. On the other hand, if the robot collects information about the user that would otherwise have been collected by an informal caregiver and the informal caregiver ultimately does not have access, they may feel replaced.

This is undoubtedly a good example to show the need not to evaluate technologies only in terms of the purposes for which they have been designed. The robot can certainly collect information perfectly well – in this sense, it does what it is supposed to. However, at the same time, it may be performing a care scenario in which the informal caregiver becomes irrelevant. It is in this sense that some authors stress the need for the monitoring task to be designed to be sensitive to the perceptions of informal caregivers. Thus, it is insisted that the robot task should be carried out in collaboration with informal caregivers rather than in competition with them (Jenkins and Draper, 2015). Besides, perhaps the problem lies precisely in the fact that the robot is effective in the task of gathering information. Should a robot designed to provide companionship collect information? The benefits associated with the interaction with the robot could be compromised if the user feels that he or she cannot trust the robot. Again, discerning which behaviour the robot prescribes to the user seems to be a fundamental task here.

In the process of patient data collection, besides privacy issues, it is relevant to consider other dynamics accompanying 'datafication' that occur in the network where the robot is embedded. Some studies on telemedicine, technologies often embedded in care robots, warn that such technological solutions based on data collection shrink and parcel out the patient, while at the same time dividing the clinical skills of medical staff (Mort, May and Williams, 2003). This trend is grounded in a process of fragmentation of care involving digitalization, an issue that raises important questions about the social value of care and the living conditions of caregivers, especially women (Vallès-Peris and Domènech, 2020a).

The process of robots' introduction in hospitals during COVID-19 is a good example to show the transcendence of a debate regarding fragmentation. These are robots designed to perform disinfection, sterilization and quarantine facilitation tasks (Khan, Siddique and Lee, 2020). Such devices use ultraviolet radiation, which they adjust according to the detection of microbes, and integrate motion sensors to avoid irradiating humans. These robots could be seen as performing care-related activities, as anything to do with cleaning is traditionally considered a care activity (Fisher and Tronto, 1990; van Wynsberghe, 2015). In the process of fragmentation that accompanies the automation of care, care tasks are organized hierarchically, with the tasks that are currently performed by cleaners being given so little consideration that they even disappear from the debate. This omission has relevant ethical, social and political consequences for how the debate about robotics is articulated, making certain contexts and care relations invisible as well as removing the voice of the people involved in those relations. What about, for example, the technological competencies of female hospital cleaners? How might their working conditions and health be harmed or benefited by the introduction of these devices? What forms of relationships and interactions does the hospital prioritize with COVID-19 patients, and how do these relationships intervene in the process of recovery and health of the sick? (Vallès-Peris, 2021).

The design process

In 2016, MedCity News, an online news source for the business of innovation in healthcare, titled one of its articles: 'As robotic design evolves, one challenge is how to make them better accepted by people' (Baum, 2016). This illustrates very well the role that designers and engineers often attribute to social scientists: that of helping them to get people to like their technological innovations. However, if one considers contributions from STS, such as those shown so far, it becomes clear that the role of social scientists should be contemplated much earlier than the quote suggests.

Indeed, if we take into account the considerations advanced so far, it becomes evident that the ethical debate should be considered during the design phase of robots, and not just when they have already been commercialized and incorporated into society (Jenkins and Draper, 2015; Iosa et al, 2016; Matsuzaki and Lindemann, 2016). This is particularly relevant when it comes to robots intended for human care. As Borenstein and Pearson (2010) point out, given that during the design process decisions have to be made about the different possibilities that are opened up, it is necessary to realize the moral burden of each of these decisions. In this sense, for these authors, the role of ethicists should not be to provide a specific design template but to help shape the values that influence the decisions of the design community to include certain features and not others.

Of particular interest in this regard is the work of van Wynsberghe (2013) and her reflection on how care robots can be designed in such a way that supports and promotes the fundamental values of care. Based on approaches aligned with the thesis that technology is not neutral, this author applies value-sensitive design combined with the care ethics perspective to incorporate ethics into the design process. Influenced by the work of both Verbeek (2006) and van Wynsberghe (2013), this perspective justifies the importance of attending to the design issue in the case of care robots, based on three arguments: (1) there are no guidelines or standards for robot design outside the factory; (2) the lack of an attempt to facilitate a way in which ethics may be translated for engineers/designers; (3) the co-shaping of technologies and societal values and norms on the development of technology.

Reflections such as these, and to ensure that ethical issues are considered, have also led to consideration of the need to involve potential users in robot design processes. Until now, the opinion of the people involved in the use of robots was rarely considered. Some groups were even more under-represented than others, as pointed out by Jenkins and Draper (2015), in terms of the interests of people caring for the elderly. This aspect is interesting because caregiving robots are often envisioned as devices aimed at caregivers. However, as Borenstein and Pearson (2010) make clear, if ethically responsible designs are promoted, care robots could be beneficial not only for the people receiving care but also for those providing care. The well-being of the caregiver is also affected by the introduction of a care robot.

In the case of older people, as Lazar et al (2016) found, although they are sometimes considered, they are not consulted; at most, in the usability and evaluation phases, but rarely in the design phase, through participatory processes. They even report a case in which, even though older people were involved in a participatory process, their preferences were ignored by the designers.

Conclusion

To conclude, and by way of summary, let us recall the main ideas that have been developed here.

Much emphasis should be placed on the idea that robots are not neutral (Sparrow, 2016; van der Meulen and Bruinsma, 2019). Every technological innovation, and a social robot is no exception, incorporates values, ways of seeing the world and promises of future societies. Incorporating social robots into care practices is neither good nor bad in itself, but neither is it innocuous. It will have effects on the way of caring, insofar as it will transform relationships, affect people's identities and change the meanings of certain situations. It is therefore essential, when considering the use of social robots for caregiving, to ask ourselves questions about issues such as the values that will be strengthened, the social relationships that will be made possible and those that will be hindered, and how the incorporation of a given social robot can improve caregiving.

Once it is assumed that technology is not neutral, in becomes imperative to unpack the discussion about ethics and the values that technologies embed. This debate, however, must be approached in a way that avoids the dominant idea that technologies should be morally evaluated only in terms of the goals for which they are designed or the quality of their performance. The focus should be on the prescriptive capacity of technology. Besides, the reflection should be developed on a case-by-case basis, based on real and/or plausible experiences. Debates inspired by futuristic fantasies can contribute little to calm and reasoned decision making.

Taking the idea of the robot embedded in a network as a reference, it is possible to find a framework for discussion, given that it is within this framework of everyday, particular and specific care practices that the problems and rules for resolving these problems emerge. On the contrary, when the debate revolves around major risks or hypothetical technological developments, some situations and conflicts are overstated, while others go unnoticed. If reflection focuses only on a utopian or dystopian development of robotics, the contexts, situations and controversies surrounding care robots that do not correspond to this imaginary future go unnoticed, as well as its effects on our lives and in the practices of care.

Robots for the common good will be those robots that incorporate processes and dynamics in which 'good care' is at the centre of the debate, with responsible robotics, at the service of individual and collective good life, integrating democratic criteria. So, the process of designing and developing robots is reconfigured as an open process of conflict and discussion, in which diverse actors negotiate around the organization and practices of care. In this sense, it has been proposed to extend citizen involvement beyond the evaluative role as target users of innovations. Too often, innovators turn to social scientists to help them make their technological proposals more acceptable to the public. This shows that some innovations are conceived on the basis of what can be considered to be a basic error: innovate for the people, but without the people. The best way to avoid people's rejection of technological innovations is to make them part of the development process. If robots are to be developed, it is essential to involve in the process the different actors who, in one way or another, are involved in their use (patients, relatives, caregivers, health professionals and so on) and to find ways to enable their collaboration with engineers and designers.

And a final consideration. Care is a very important part of our lives and, therefore, any transformation of it should be approached with prudence and caution. The way in which we care says a lot about the kind of society we are, so proposing substantial changes in the way we carry it out is not a minor affair. Precisely because of the importance we attach to care, all too often the debates surrounding the possibility of introducing social robots easily resonate with other very simplistic discussions around the fear that a cold technology could be implemented at the cost of warm human care (Pols and Moser, 2009).

At this stage, it is important to recognize that we must deal with the uncertainty and that we do not know the real impact that the introduction of robots in care could have. In this scenario of uncertainty, social sciences can provide some ideas about the best way to deal with it. In this regard, the work of Callon, Lascoumes and Barthe (2001) is particularly relevant. What these authors propose is that, in situations of uncertainty, it is as unwise to act without a clear idea of the repercussions of the action as it is to plunge into inactivity and paralysis. It is *measured action* that is required in these types of circumstances.

We have explained elsewhere (Vallès-Peris and Domènech, 2020b) how the proposal of measured action could be applied to the case of social robots. To conclude this chapter, let us recall some of these implications. On the one hand, it is important to note that any process of introducing care robots must start from an awareness of the fears and hopes of the people involved. On the other hand, such introductory processes should contain the possibility of rectifying the direction taken. In this sense, social scientists can provide the information needed to determine when a change of course is necessary. Finally, when experiences are developed, continuous monitoring of the outcomes that become apparent must be carried out in order to make sure that the effects produced are the desired ones. If the social sciences manage to play a relevant role in the development of such tasks, it will not be necessary for engineers and designers to call on them to find a way to make social robots acceptable to the public.

References

Akrich, M. (1992) 'The de-scription of technical objects', in W.E. Bijker and J. Law (eds) *Shaping Technology/Building Society, Studies in Sociotechnical Change* (3rd edn), Cambridge, MA: MIT Press.

Baum, S. (2016) 'As robotic design evolves, one challenge is how to make them better accepted by people', *MedCityNews*, 12 March, https://medc itynews.com/2016/03/robot-and-human-interactions/.

Belanche, D. et al (2020) 'Service robot implementation: a theoretical framework and research agenda', *Service Industries Journal*, 40(3–4): 203–225.

Bijker, W.E. (2009) 'How is technology made? That is the question!' *Cambridge Journal of Economics*, 34(1): 63–76.

Borenstein, J. and Pearson, Y. (2010) 'Robot caregivers: harbingers of expanded freedom for all?' *Ethics and Information Technology*, 12(3): 277–288.

Callon, M. (1987) 'Society in the making: the study of technology as a tool for socio- logical analysis', in W. Bijker, T. Hughes and T. Pinch (eds) *The Social Construction of Technological Systems: New Directions in the Sociology and History of Technology* (4th edn), London: MIT Press, pp 83–103.

Callon, M. and Latour, B. (1991) 'Introduction', in M. Callon, and B. Latour (eds) *La science telle qu'elle se fait* , Paris: La découverte, pp 7–36.

Callon, M., Lascoumes, P. and Barthe, Y. (2009) *Acting in an Uncertain World: An Essay on Technical Democracy,* Cambridge, MA: MIT Press.

Cardona, M., Cortez, F., Palacios, A. and Cerros, K. (2020) 'Mobile robots application against COVID-19 pandemic', *IEEE ANDESCON*, 1–5.

Coeckelbergh, M., Pop, C., Simut, R., Peca, A., Pintea, S., David, D. et al (2016) 'A survey of expectations about the role of robots in robot-assisted therapy for children with ASD: ethical acceptability, trust, sociability, appearance, and attachment', *Science and Engineering Ethics*, 22(1): 47–65.

Díaz, P.M. (2017) 'Japón confía en la última generación de robots para cuidar a los ancianos', ABC, 1 October, www.abc.es/sociedad/abci-japon-con fia-tercera-generacion-robots-para-cuidar-ancianos-201709302021_noti cia.html.

Díaz-Boladeras, M., Angulo, C., Domènech, M., Albo-Canals, J., Serrallonga, N., Raya, C. et al (2016) 'Assessing pediatrics patients' psychological states from biomedical signals in a cloud of social robots', in E. Kyriacou, S. Christofides and C.S. Pattichis (eds) *XIV Mediterranean Conference on Medical and Biological Engineering and Computing, MEDICON 2016, Paphos, Cyprus. IFMBE Proceedings*, Springer International Publishing, pp 1179–1184.

Ding, M., Ikeura, R., Mori, Y., Mukai, T. and Hosoe, S. (2013) 'Measurement of human body stiffness for lifting-up motion generation using nursing-care assistant robot – RIBA', 2013 *IEEE Sensors*, 1–4, https://doi.org/10.1109/ICSENS.2013.6688431.

Duch-Brown, N., Rossetti, F. and Haarburger, R. (2021) *Evolution of the EU Market Share of Robotics: Data and Methodology, EUR 30896 EN*, Luxembourg: Publications Office of the European Union.

Feenberg, A. (2010) *Between Reason and Experience: Essays in Technology and Modernity*, Cambridge, MA: MIT Press.

Feil-Seifer, B.D. and Matari, M.J. (2011) 'Socially assistive robotics: ethical issues related to technology', *Robotics Automation Magazine*, 18(1): 24–31.

Fine, M. and Tronto, J. (2020) 'Care goes viral: care theory and research confront the global COVID-19 pandemic', *International Journal of Care and Caring*, 4(3): 301–309.

Fisher, B. and Tronto, J. (1990) 'Toward a Feminist Theory of Caring', in E.K. Abel and M.K. Nelson (eds) *Circles of Care: Work and Identity in Women's Lives*, New York: SUNY Press, pp 35–62.

Foster, M. (2018) 'Aging Japan: robots may have role in future of elder care', Reuters, 28 March, www.reuters.com/article/us-japan-ageing-robots-wid erimage-idUSKBN1H33AB.

González-González, C.S., Violant-Holz, V. and Gil-Iranzo, R.M. (2021) 'Social robots in hospitals: a systematic review', *Applied Sciences*, 11(13): 5976.

Heerink, M., Vanderborght, B., Broekens, J. and Albó-Canals, J. (2016) 'New friends: social robots in therapy and education', *International Journal of Social Robotics*, 8(4): 443–444.

Hung, L., Gregorio, M., Mann, J., Wallsworth, C., Horne, N., Berndt, A. et al (2021) 'Exploring the perceptions of people with dementia about the social robot PARO in a hospital setting', *Dementia*, 20(2): 485–504.

IFR (2021) 'IFR presents World Robotics 2021 reports', IFR Press Room, 28 October, https://ifr.org/ifr-press-releases/news/robot-sales-rise-again.

Iosa, M., Morone, G., Cherubini, A. and Paolucci, S. (2016) 'The three laws of neurorobotics: a review on what neurorehabilitation robots should do for patients and clinicians', *Journal of Medical and Biological Engineering*, 36(1): 1–11.

Jenkins, S. and Draper, H. (2015) 'Care, monitoring, and companionship: views on care robots from older people and their carers', *International Journal of Social Robotics*, 7(5): 673–683.

Jokela, M. (2019) 'Patterns of precarious employment in a female-dominated sector in five welfare states – the case of paid domestic labor sector', *Social Politics*, 26(1): 116–138.

Khan, Z.H., Siddique, A. and Lee, C.W. (2020) 'Robotics utilization for healthcare digitization in global COVID-19 management', *International Journal of Environmental Research and Public Health*, 17(11): 3819.

Khosla, R., Chu, M.-T., Sadegh Khaksar, S.M., Nguyen, K. and Nishida, T. (2021) 'Engagement and experience of older people with socially assistive robots in home care', *Assistive Technology*, 33(2): 57–71.

Klairmont, L. (2017) 'Kids isolated by cancer connect online', CNN, 1 December, https://edition.cnn.com/2017/03/16/health/cnn-hero-les lie-morissette-grahamtastic-connection/index.html.

Kristoffersson, A., Coradeschi, S. and Loutfi, A. (2013) 'A review of mobile robotic telepresence', *Advances in Human-Computer Interaction*, 2013: 902316.

Kritikos, M. (2020) 'Ten technologies to fight coronavirus', European Parliament, European Parliamentary Research Service.

Larriba, F. (2016) 'Externalising moods and psychological states in a cloud-based system to enhance a pet-robot and child's interaction', *BioMedical Engineering Online*, 15(1): 187–196.

Lässig, R. (2021) 'Robotics outlook 2030: how intelligence and mobility will shape the future', Boston Consulting Group, 28 June, www.bcg.com/publications/2021/how-intelligence-and-mobility-will-shape-the-future-of-the-robotics-industry.

Latour, B. (1994) 'On technical mediation', *Common Knowledge,* 3(2): 29–64.

Latour, B. (2005) *Reassembling the Social: An Introduction to Actor-Network Theory*, Oxford: Oxford University Press.

Law, J. and Mol, A. (1995) 'Notes on materiality and sociality', *The Sociological Review*, 43(2): 274–294.

Lazar, A., Thompson, H.J., Piper, A.M. and Demiris, G. (2016) 'Rethinking the design of robotic pets for older adults', in *Proceedings of the 2016 ACM Conference on Designing Interactive Systems*, pp 1034–1046.

Lightman, N. (2022) 'Comparing care regimes: worker characteristics and wage penalties in the global care chain', *Social Politics: International Studies in Gender, State & Society*, 28(4): 971–998.

Lipp, B. (2022) 'Caring for robots: how care comes to matter in human–machine interfacing', *Social Studies of Science*, https://doi.org/10.1177/03063127221081446.

Magnenat-Thalmann, N. and Zhang, Z. (2014) 'Assistive social robots for people with special needs', in *2014 International Conference on Contemporary Computing and Informatics (IC3I)*, pp 1374–1380.

Maibaum, A., Bischof, A., Hergesell, J. and Lipp, B. (2021) 'A critique of robotics in health care', *AI and Society*, 37: 467–477.

Makedon, V., Mykhailenko, O. and Vazov, R. (2021) 'Dominants and features of growth of the world market of robotics', *European Journal of Management Issues*, 29(3): 133–141.

Matsuzaki, H. and Lindemann, G. (2016) 'The autonomy-safety-paradox of service robotics in Europe and Japan: a comparative analysis', *AI and Society*, 31(4): 501–517.

Miller, B. (2021) 'Is technology value-neutral?' *Science Technology and Human Values*, 46(1): 53–80.

Mol, A., Moser, I. and Pols, J. (2015) *Care in Practice: On Tinkering in Clinics, Homes and Farms*, Bielefeld: Transcript Verlag.

Mort, M., May, C.R. and Williams, T. (2003) 'Remote doctors and absent patients: acting at a distance in telemedicine?', *Science, Technology, & Human Values*, 28(2): 274–295.

Murphy, R.R., Gandudi, V.B.M. and Adams, J. (2020) 'Applications of robots for COVID-19 response', *arXiv*, https://doi.org/10.48550/arXiv.2008.06976.

Nickelsen, N.C.M. and Simonsen Abildgaard, J. (2022) 'The entwinement of policy, design and care scripts: providing alternative choice-dependency situations with care robots', *Sociology of Health & Illness*, 44(2): 451–468.

Papadopoulos, I., Koulouglioti, C. and Ali, S. (2018) 'Views of nurses and other health and social care workers on the use of assistive humanoid and animal-like robots in health and social care: a scoping review', *Contemporary Nurse*, 54(4–5): 425–442.

Pols, J. and Moser, I. (2009) 'Cold technologies versus warm care? On affective and social relations with and through care technologies', *ALTER – European Journal of Disability Research / Revue Européenne de Recherche sur le Handicap*, 3(2): 159–178.

Redacción QMayor (2016) 'Un nuevo sistema robótico puede ayudar a personas mayores y con discapacidad', *QMayor Magazine*, 18 June, www.qmayor.com/salud/personas-mayores-y-discapacidad/.

Rommetveit, K., van Dijk, N. and Gunnarsdóttir, K. (2020) 'Make way for the robots! Human- and machine-centricity in constituting a European public–private partnership', *Minerva*, 58(1): 47–69.

Roser, M. (2017) 'Link between health spending and life expectancy: US is an outlier', in *Our World in Data*, https://ourworldindata.org/the-link-between- life-expectancy-and-health-spending-us-focus.

Royakkers, L. and van Est, R. (2015) 'A literature review on new robotics: automation from love to war', *International Journal of Social Robotics*, 7(5): 549–570.

Šabanović, S. and Chang, W.L. (2016) 'Socializing robots: constructing robotic sociality in the design and use of the assistive robot PARO', *AI and Society*, 31(4): 537–551.

Savela, N., Turja, T. and Oksanen, A. (2018) 'Social acceptance of robots in different occupational fields: a systematic literature review', *International Journal of Social Robotics*, 10(4): 493–502.

Schaal, S. (2007) 'The new robotics – towards human-centered machines', *HFSP Journal*, 1(2): 115–126.

Schillmeier, M. and Domenech, M. (eds) (2010) *New Technologies and Emerging Spaces of Care*, Farnham: Ashgate.

Sharkey, A. and Sharkey, N. (2011) 'Children, the elderly, and interactive robots: Anthropomorphism and deception in robot care and companionship', *IEEE Robotics and Automation Magazine*, 18(1): 32–38.

Sparrow, R. (2016) 'Robots in aged care: a dystopian future?' *AI and Society*, 31(4): 445–454.

Sparrow, R. and Sparrow, L. (2006) 'In the hands of machines? The future of aged care', *Minds and Machines*, 16(2): 141–161.

Topol, E.J. (2019) 'High-performance medicine: the convergence of human and artificial intelligence', *Nature Medicine*, 25(1): 44–56.

Vallès-Peris, N. (2021) 'Repensar la robótica y la inteligencia artificial desde la ética de los cuidados', *Teknokultura. Revista de Cultura Digital y Movimientos Sociales*, 18(2): 137–146.

Vallès-Peris, N. and Domènech, M. (2020a) 'Roboticists' imaginaries of robots for care: the radical imaginary as a tool for an ethical discussion', *Engineering Studies*, 12(3): 157–176.

Vallès-Peris, N. and Domènech, M. (2020b) 'Robots para los cuidados. La ética de la acción mesurada frente a la incertidumbre', *Cuadernos de Bioética*, 31(101): 87–100.

Vallès-Peris, N. and Domènech, M. (2021) 'Caring in the in-between: a proposal to introduce responsible AI and robotics to healthcare', *AI & Society*, https://ddd.uab.cat/record/251084.

Vallès-Peris, N., Angulo, C. and Domènech, M. (2018) 'Children's Imaginaries of Human-Robot Interaction in Healthcare', *Environmental Research and Public Health*, 15: 970.

Vallès-Peris, N., Barat-Auleda, O. and Domènech, M. (2021) 'Robots in healthcare? What patients say', *International Journal of Environmental Research and Public Health,* 18: 9933.

van der Meulen, S. and Bruinsma, M. (2019) 'Man as "aggregate of data": what computers shouldn't do', *AI & Society*, 34(2): 343–354.

van der Plas, A., Smits, M. and Wehrmann, C. (2010) 'Beyond speculative robot ethics: a vision assessment study on the future of the robotic caretaker', *Accountability in Research*, 17(6): 299–315.

van Wynsberghe, A. (2013) 'Designing robots for care: care centered value-sensitive design', *Science and Engineering Ethics*, 19(2): 407–433.

van Wynsberghe, A. (2015) *Healthcare Robots. Ethics, Design and Implementation*, Abingdon: Routledge.

Verbeek, P.-P. (2006) 'Materializing morality: design ethics and technological mediation', *Science, Technology & Human Values*, 31(3): 361–380.

WHO (2021) 'Ageing and health', WHO Newsroom, 4 October, www. who.int/news-room/fact-sheets/detail/ageing-and-health.

WHO (2021b) 'Dementia', WHO Newsroom, 2 September, www.who. int/news-room/fact-sheets/detail/dementia.

9

Are Ovulation Biosensors Feminist Technologies?

Joann Wilkinson and Celia Roberts

Introduction

Towards the end of 2019, media reports emerged of a team of student researchers at the University of Copenhagen developing a chewing gum that helps women detect ovulation: when the 'Ovulaid' gum changes colour, women will know they've reached their 'fertile window', the days of the menstrual cycle when a woman can become pregnant (Russell, 2019). Ovulation, the releasing of an egg from the ovaries, is triggered by the rise and fall of the so-called 'sex' hormones, oestrogen, progesterone and luteinizing hormone. These hormonal changes can (sometimes) be detected through saliva, urine and temperature. Ovulation biosensors, like the chewing gum, claim to help women detect and predict their ovulation patterns and thus make conception easier by indicating the times at which they are most likely to conceive (Wilkinson, 2016). They can also, of course, be used to avoid conception.

Ovulation biosensors are located within a wider array of 'FemTech' devices/products comprising an ever-increasing number of fertility-related apps and devices that have come onto the global market, each attempting to offer something new such as more in-depth data or greater levels of accuracy. Media stories, like the one cited, frequently accompany these developments, focusing not only on the increasing use of such devices (see, for example, Weigel, 2016) but also on the poor reliability of fertility tracking (see, for example, Kleinman, 2021). The FemTech industry (services, products and software designed to focus on women's health) is increasingly subjected to public scrutiny. Many self-tracking apps and devices encourage women to collect and store large quantities of bodily data online, and controversies have arisen around companies' capture and on-selling of this data (see, for example, Shadwell, 2019). In an important case, in 2021 Flo Health Inc, (owners of 'Flo', a period and ovulation tracking app with more than 100 million users), settled with the US Federal Trade Commission, who proved the company was sharing users' data with third parties, including Google and Facebook, despite having promised to keep it secure (Federal Trade Commission, 2021).

FemTech is also increasingly the subject of social research, as we describe later in this chapter. A key strand of this work highlights issues of data privacy and the commercialization of women's fertility. Science and technology studies (STS) and feminist technoscience studies (FTS) approaches are central to this field, and we situate our research in this space. Since 2011, we have been part of a wider research team initially based at Lancaster University in the North West of England called the Living Data Research Group. Funded by Intel Labs' University Research Office's Biosensors in Everyday Life programme, Joann undertook a PhD, supervised by Celia and Maggie Mort, on ovulation biosensing (Wilkinson, 2016). This ethnographic research was part of a group of STS and FTS studies analysing biosensing across the lifecourse (Roberts et al, 2019). Here, we draw upon the work of this group and others in the now-burgeoning field of social studies of ovulation and menstrual tracking to explore the connections between these technologies and feminist health politics to ask: are ovulation biosensors feminist technologies? In the light of public and academic criticism of data on-selling and the capitalization of fertility (Roberts and Waldby, 2021), we investigate how women engage with ovulation biosensors and explore what it means to know and do ovulation biosensing. We also ask: how do these practices build on earlier feminist experiments around women's health, bodies and the reshaping of science?

Joann's study, which took place between 2011 and 2016, focused on cis-women in heterosexual relationships wanting to have a child. Some of the women had only just started trying to conceive, while others had been trying for longer than 12 months. Our methods included 29 face-to-face interviews, analysis of women's participation on online fertility forums and textual analysis of manufacturers' promotional materials such as websites, video and leaflets. In this chapter, we present data from women's discussions on online fertility forums and interviews with women trying to conceive. Ethical approval was granted by Lancaster University's Ethics Committee (Wilkinson et al, 2015; Wilkinson, 2016).

What do ovulation biosensors do?

Biosensors are defined by Intel anthropologist Dawn Nafus (2016, p xiii), the Director of the Biosensors in Everyday Life programme, as devices that indicate something about the body or physical environment, and biosensing practice as the use of information technology to understand something about the body or environment. A plethora of biosensing or self-tracking tools, both high- and low-tech, are currently available on the market, to facilitate monitoring of weight, health, sleep, diet, mood or menstruation, for example (FitBits and Apple Watches are perhaps the two best known). A growing body of research explores the social aspects of biosensing and self-tracking, focusing on practices of consumption (Lupton, 2014a; Davies, 2015; Harris et al, 2016), surveillance (Kenner, 2008), the regulation of personal medical technologies (Faulkner, 2009; Pantzar and Ruckenstein, 2015), how these are used in conjunction with social media (Rettberg, 2014) or the connection between knowledge and action (Neff and Nafus, 2016). Our group's book, *Living Data: Making Sense of Health Biosensing* (Roberts et al, 2019), describes ethnographies of various biosensing practices across the lifecourse, arguing that these must be understood within wider contexts of everyday lives, cultural logics and technical platforms. In particular, the book contests the popular claim that 'more data' leads to 'better health'.

It is important to understand the histories of particular biosensing practices. Although the FemTech industry aims to convince us that we are on the cusp of a revolution in birth control methods (Savage, 2017), ovulation tracking has in fact been around for some time. In the UK in the 1930s, doctors recorded ovulation to observe the effects of infertility drugs (Chen and Wallach, 1994). In the 1960s, a form of ovulation tracking known as 'The Billings Method' was promoted by the Roman Catholic Church as a morally acceptable form of birth control and as an alternative to the contraceptive pill. With the introduction of new reproductive technologies in the 1970s and 1980s, such as in-vitro fertilization (IVF) and intrauterine insemination (ICI), ovulation was identified as important for patients to understand. It was not long before practices of recording ovulation moved out of the clinic and into the domestic sphere, with the first home ovulation testing kit coming onto the market in the 1980s. Although uptake by home/domestic users was initially slow, since the turn of the century use of domestic ovulation tracking technologies has increased exponentially, as forms of sensing and recording have proliferated (Wilkinson, 2016).

Although contemporary ovulation biosensors and fertility apps overlap in some ways, there are some key differences. Ovulation biosensors 'sense' bodily substance or quality (urine, saliva or temperature) in order to detect hormonal changes. Although some are linked to electronic devices, apps and online platforms can also be materially disconnected from them – that is, they can exist as stand-alone devices. Fertility tracking apps, in contrast, function like 'living' diaries, requiring users to contribute a considerable amount of personal data by inputting, for example, dates of menstruation, sexual intercourse, symptoms of ovulation and menstruation, mood, diet and sleep. The boundaries between biosensors and apps are sometimes blurred as manufacturers develop devices that allow apps to 'read' home testing kits and to use this data in their predictive algorithms.

The ovulation biosensors Joann studied in her PhD include the ovulation microscope, ovulation predictor kits and the basal body temperature thermometer. The ovulation microscope is a small cylindrical device resembling a lipstick tube. Like the Danish chewing gum, this device can purportedly detect changing levels of oestrogen in saliva leading up to ovulation. The user places a small amount of saliva onto the lens and then peers through the microscope's 'eye'. If oestrogen is present in sufficient quantities, crystallized ferning patterns will appear. In the UK, the device can be bought online for approximately £20 and is reusable.[1]

Another commonly used biosensor is single-use testing strips, also known as ovulation predictor kits (OPKs) or dipsticks. They are small strips of paper which, similar to pregnancy tests, are dipped in urine to test for a luteinizing hormone surge that typically takes place shortly before the egg is released from the ovary. If the test is positive, the line will be darker or as dark as the control line. Ovulation testing strips can be purchased in quantities of 20, 50 or 100 through online retailers who specialize in healthcare products. On online fertility forums, women refer to these as 'cheapies' because of their low cost in comparison to branded ovulation kits – in the UK, a pack of 100 tests costs around £10.[2]

The basal body temperature refers to the lowest temperature that can be recorded in the body after a period of prolonged rest. In a third form of ovulation biosensing, women use a basal body thermometer, which detects small differences in temperature, to take their temperature immediately on waking, physically marking this on a chart over the menstrual cycle.[3]

This method is based on the understanding that the basal body temperature during the first half of a woman's cycle will be lower than the second half, and ovulation is the moment of change between the two parts of the cycle, observed through a 'dip and sharp rise'. However, the basal body temperature method does not give women advance warning of when ovulation will take place. Instead, it is often used to confirm ovulation or to give a general idea of when this takes place each month. The thermometer can be bought online for between £3 and £10 and comes accompanied by a small number of charts which can also be downloaded from fertility, pregnancy and parenting websites.

Since Joann's study, a new generation of fertility-monitoring technologies have appeared. Some of these are stand-alone apps such as *Glow* or *Clue* that allow users to track different aspects of their reproductive lives including period data, fertility, pregnancy and sexual activity. These apps require users to input a wide range of personal data as a way to 'illuminate health through data and empower people with new information about their bodies' (Glow,

2021). However, an increasing number of apps are connected to biosensors. For example, *Natural Cycles*, the first digital method of birth control to be regulated by the US Food and Drug Administration (FDA), brings together a smartphone app with a basal body temperature thermometer. Women take their temperature on waking and input this and other data, such as length of cycle and period data, into the app. The app then informs them if it is a 'Green Day' (not fertile) or 'Red' (use protection). The *Inito* fertility hormone tracker, which claims to measure three hormones at the same time (oestrogen, luteinizing hormone and progesterone), goes one step further in hybrid tracking by providing urine-based dipsticks and a 'reader' that the user attaches to their phone so that test results can be transmitted directly to the app platform. This displaces the role of the user in reading the data and may prevent women learning about ovulation in the ways we discuss later. Manufacturers of fertility-tracking devices, it seems, are searching for ways to produce seamless data transmission between bodies and algorithmic platforms.

Feminism and ovulation tracking

Ovulation biosensing practices aim to address the temporal gap that emerges between trying to become pregnant and conception. This time is fraught with tensions and excitement as women and/or couples invest physically and emotionally in creating a pregnancy but do not know how long this may take or if they will succeed. Ovulation biosensors feed into public health debates around individuals taking responsibility for their bodies by accessing the right kinds of data to achieve the best outcomes. As Roberts et al argue, 'good citizens are expected to know and understand ovulation and menstruation' (Roberts et al, 2019, pp 36–37).

Women learning about their own bodies has long been a focus of feminist health movements. Rooted in the women's liberation movements of the 1960s and 1970s, the central 'protocol' (Murphy, 2012) of women's health movements was the collective investigation of bodies, relationships, sexuality and health in order to form an evidence base for political action. STS scholar Michelle Murphy (2004, p 347) argues that 'The central epistemological principle of feminist self-help, as with radical feminism more generally, was that all knowledge production should begin with women's experiences.' The creation of feminist self-help clinics – that is, groups of women coming together to share experiences and learn about their bodies through direct observation – constituted a 'mobile set of practices, a mode for arranging knowledge production' (Murphy, 2012, p 25). Such feminist protocols established ' "how to" do something, how to compose the technologies, subjects, exchanges, affects, processes, and so on that make up a movement of health care practice' (Murphy, 2012, p 25). Feminist standpoint theory and the works of Dorothy Smith (1990), Sandra Harding (1986) and Donna Haraway (1997), among others, followed on from these feminist practices, highlighting the significance of lived experience and structural positionality in knowledge production.

Meetings included collective visual examinations of genitals and cervixes, and feeling the size and position of women's uteruses. The idea was that women could conduct these practices in their own domestic spaces using common items from the home. Conventionally, feminists argued, female bodies were controlled by the paternalistic and judgemental scrutiny of male medical authority (Kline, 2010, p 14): new knowledges and practices were, therefore, the only route forward. In the US, the Boston Women's Health Book Collective, for example, established a set of bodily and writing practices to produce the book, *Our Bodies Ourselves: A Book by and for Women* (1971), which became a best-seller that was republished over many decades and is now a detailed website. 'By incorporating personal experience into the narrative', historian Wendy Kline argues, '[women] began the process of transforming medical knowledge into something subjective, political and empowering' (Kline, 2010, p 15).

Fertility was a key theme in this work, since women's health movements had a strong focus on contraception, abortion, pregnancy and birth. Since the 1980s, discourses around (in)fertilities have also become increasingly visible within public health, medical and lay contexts, emerging as a legitimate area of concern and interest. Although it is recognized that fertility decreases over the lifecourse (Faddy and Godsen, 2003; Balasch and Gratacós, 2012), age-related infertility is often positioned as a medical condition, and one which can or *should be* avoided: we hear much discussion in the British and Australian media about the 'optimum time' to become pregnant (Hope, 2013) or of motherhood as 'delayed', thereby locating fertility within a framework of risk and choice (Campbell, 2014). Through such discussions, women are taught to expect problems with fertility, indeed to 'anticipate infertility' (Martin, 2010). The underlying message is that female fertility is both a 'precious' and time-bound resource for which women have a responsibility to care (Roberts and Waldby, 2021). Feeding into these debates is the understanding that individuals can, and should, take responsibility for their own health concerns and that greater quantities of information, data and pioneering technoscience will allow individuals to do this. The implication is that if women want to become pregnant, there is no reason for them to struggle; by using the appropriate devices, such as ovulation biosensors, or accessing the right kinds of data, conception can be made easier.

An extensive body of feminist scholarship examines the biomedicalization of in/fertility, including studies of IVF (Franklin, 1997; Throsby, 2004; Thompson, 2005), pre-implantation genetic diagnosis, (Franklin and Roberts, 2006), egg freezing (Waldby, 2015), egg donation (Ragone, 1994;

Nahman, 2011) and amniocentesis (Rapp, 1999). A smaller number of studies have focused on how less-invasive technologies, including cheap, low-tech devices used in the home, have shaped women's experiences of conception and contraception (Mamo, 2007; Layne, 2009; Nordqvist, 2011; Lupton, 2014b). Such 'home reproductive technologies' are sometimes viewed as feminist devices. Historian Sarah Leavitt (2006), for example, argues that the home pregnancy test was revolutionary: 'removing the moment of pregnancy diagnosis from the institutional gaze of the doctor to the private gaze of the pregnant (or not-pregnant) woman herself. It is an example of the way in which the women's health movement worked to recapture women's control over much information related to pregnancy' (Leavitt, 2006, p 344). North American anthropologist Linda Layne, however, questions this view, arguing that prior to the test's inception, women were always the first to know about their bodies, albeit through other means (Layne, 2009, p 71, see also Duden, 1992). For Layne, devices such as the home pregnancy test devalue the knowledge that women have about their own bodies, encouraging them to spend money on unnecessary products, creating a 'pharma-technological dependency' (Layne, 2009, p 61), as well as profits for corporations.

There is also an increasing academic feminist interest in the use of fertility-tracking apps, focusing on: women's experiences of using them (Gambier-Ross et al, 2018; Grenfell et al, 2021); how they configure the pre-pregnant reproductive body (Hamper, 2020); and the ways in which they further medicalize women's bodies (Healy, 2021). Although it is sometimes argued that tracking apps can empower users to monitor and manage their own health (Lupton, 2014b), several studies have shown this is not the case for all users (Gambier-Ross et al, 2018). Sociological research has also demonstrated that many people give up self-tracking only a few weeks after starting, for a variety of reasons (Ledger and McCaffrey, 2014: Nafus and Sherman, 2014). Feminist research on fertility apps also raises concerns about the capture and reselling of women's data to industries involved in personalized marketing, as noted earlier (Roberts et al, 2019, p 54–63). As Roberts and Waldby (2021) argue, such profiteering is part of a much broader commodification of female fertility involving reproductive medical services such as egg freezing, gamete donation and surrogacy. There is less work to date on ovulation biosensing, perhaps because the devices are either low-tech and domestic, or rather niche, involving more expensive devices linked to 'experts' (see Wilkinson, 2016). However, we argue that ovulation biosensing deserves greater feminist and STS/ FTS attention. As Roberts et al (2019, p 26) contend, 'gadgets are … part of broader transverse relations that bring disparate groups and people into contact'. Ovulation biosensors are thus part of sociotechnical platforms reshaping fertility and our collective ideas about 'reproductive life'. The ovulation biosensing devices we discuss later are relatively low-tech and can be used without the capture and storing of data online. Addressing a gap in the literature, we want to suggest that their very domestic and practical orientations articulate clear connections to earlier practices of women's health movements and thus to feminist explorations of reproductive embodiment.

What happens when women 'do' ovulation biosensing?
Skilling up: learning how to use the tools

The women of Joann's study began using ovulation biosensors for different reasons, although all had in common the aim of becoming pregnant. Particularly for those who had avoided pregnancy for many years, tracking ovulation became part of an exciting 'journey' towards 'starting a family'. For others, ovulation biosensing signified a change in their trajectory of trying to conceive, moving from 'just trying' (having sexual intercourse without reference to fertile times), to 'actively trying' (tracking ovulation in order to time sexual intercourse). This shift suggested a greater level of effort in becoming pregnant and was often referred to as 'taking control' of conception. One of the first steps in this process was learning how to use the devices and how to make sense of the data collected. The online fertility forums were a key source of information where women could post their questions and receive advice, support and suggestions from other women trying to conceive. In the following extract from an online fertility forum, a user describes some of the difficulties she faced when using the ovulation microscope:

I have been using a ferning microscope this cycle but have been finding it hard to analyse the pattern. I am getting close to ovulation and the pattern looks different than earlier in the month but it is not a ferning pattern. It is like a criss–cross pattern, like looking at a piece of material/weave under a microscope. Does anyone know if everyone gets a ferning pattern at O? Or can they look different? Can you O without a ferning pattern? I know that I haven't ovulated yet, but am wondering when to test with an OPK [ovulation predictor kit] as I don't have many left and can't get more in a hurry. (Fertility forum user A)

I found mine really confusing at first, but what I have since realised is that the slightest turn of the microscope part makes it look totally different – in my first month I didn't see any ferning, but then in my second I thought I just had the same again, until I turned the microscope the tiniest amount, and there were all the lines ... I also found it hard at first to get the right amount of spit – as too much or too little was not giving a sensible result – so sometimes I let it dry, and then tried again if the pattern wasn't clear. (Fertility forum user B)

In this extract, user A describes the difficulties of making sense of the patterns that appear in the microscope. She tries to match this with the instruction sheet but finds that her data does not fit with the images provided showing the ferning patterns (fertile) or dots (infertile). Thus begins the task of labelling. The user draws on other descriptions to make sense of the images such as criss-crossing or weaving. Other users sometimes describe lint or spikiness; lines that are branchy or wavy; dots may be described as unconnected, partial. There may be bubbles, feathered particles, specs or empty spaces; patterns were described as good, strong or lovely. The images viewed in the microscopes extended beyond the images presented in the instruction – women's images were more complex and varied and they had to engage in practices of interpretation in order to place their bodies within some framework of meaning.

The ovulation microscope forges new connections between fertility and saliva. For many women, the relationship between saliva, hormones and fertility was unfamiliar, unlike the connections between hormones and urine-based testing (such as pregnancy tests) and blood tests. Furthermore, saliva is configured as belonging inside the body; once it travels outside, it becomes odd, dirty or unfamiliar. However, studies including one from the Living Data group on direct-to-consumer genetic testing (Kragh-Furbo, 2015), have shown saliva's growing importance; it is a 'promissory substance' that circulates commercial, scientific and social relations that can be capitalized on (Kragh-Furbo and Tutton, 2017).

In addition to being unfamiliar or odd, saliva emerges in Joann's data as a 'fragile substance' (Kragh-Furbo and Tutton, 2017), in that it has to be managed. Women must apply the right quantity (a tiny amount) and the right texture (saliva without bubbles and not too runny) on the lens, a practice that can present challenges. User B in the extract draws the other woman's attention to how she uses the biosensing device – emphasizing that ovulation can be seen only once the tool is managed correctly. In order to 'learn to see' the ferning patterns, the user must develop skills in using the microscope, such as manipulating the sample quantity and letting this dry; ensuring the sample is taken at the right time; and, finally, repeating the process when the results are not 'sensible'. Like members of women's health movements of the past, here women begin to engage in scientific practices at home (Nelson, 1977; Federation of Feminist Women's Health Centers, 1981; Boston Women's Health Book Collective, 2011).

Making sense of the data

The women in Joann's study often collected different kinds of data (that is, not only through ovulation biosensing) and struggled to discern coherence across these sets. For example, in addition to using the ovulation microscope or hormone strips, some users already had a sense of how long their ovulatory cycle should be, or when ovulation 'should' take place. They observed how the texture of cervical fluid or libido changed over the days leading up to ovulation. They sometimes made a (mental) note of ovulation pain: cramps, 'mittelschmerz', or bloating, symptoms which marked this time of their cycle. They also frequently used more than one biosensor device, for example, combining basal body temperature monitoring with hormone strips. The data that women collected often gave a conflicting picture of ovulation, as the user describes in the following post:

> We decided to try OPKs [ovulation predictor kits] this month, and I am now on CD11 [calendar day]. I started testing on CD8, as per the instructions on the pack. However, I still haven't got a positive, or rather my faces are not smiley! I only have a 25 day cycle, surely I should have ovulated by now? The method that we used with dd [darling daughter] was just to BD [baby dance] when I had the most EWCM [egg white cervical mucous] which I had loads of yesterday, and expected a positive test today, but it was negative. Now I am worried that I'm not ovulating. Can someone help explain it all to me?
> [Later]
> I have quite a lot of EWCM today so I will BD tonight. I think and still test tomorrow. I am really wondering why I bought them. I never used them when ttc [trying to conceive] dd, they are just stressing me out!! (Fertility forum user)

In the extract cited here, the user brings together different sets of data but these create an experience of ovulation that is confusing and unclear, which is stressful. Women post their concerns on online fertility forums and the responses vary considerably. Some women argue that hormone strips (OPKs) are confusing, unhelpful or 'never give a positive result', despite describing how they 'became pregnant in the end'. Others explain how they have always obtained a positive test result, but never a pregnancy. For still others, ovulation biosensors are helpful in that they show that ovulation occurs much later than expected, although for many users cervical fluids and libido become a more reliable indicator of when to engage in reproductive heterosexual intercourse.

Frequently, however, women 'probe' the data posted on online fertility forums, asking questions or making suggestions about what they could mean and how they fit into the bigger picture of ovulation and women's attempts to become pregnant. Here begins a process of collectively theorizing the data; of searching for patterns and establishing meaning. Women calibrate and recalibrate understandings of their bodies and fertilities in line with these discussions and it is through these sense-making practices that women come to 'know' their ovulation in some way. For us, there is strong resonance with the collective activities of feminist health groups (albeit without physical co-presence). In addition to this collective sensing-making, ovulation biosensing changes women's sensing practices in other ways.

Becoming the sensor

As already noted, social research has shown that people often stop using self-tracking devices after a short time, either because they have acquired the information they need or because it has become less interesting. Women's engagement with ovulation biosensing is time bound (either to the point of knowing ovulation or to the point of conception), although some may continue to use biosensors to avoid pregnancy at a later date. Joann's research shows that although women may come to know or understand ovulation in more ways they experience as more useful, after some time, it is the women who do the sensing:

'I also started to read some signs of my body as it coincided with when I knew when I was ovulating due to the strips as well, which I would never have sensed or thought about before … I could feel I was ovulating so I could feel maybe some very slight cramping on one side.' (Interview with Mel)

'Yeh I did still use them and it got to the point more recently where I could predict and I would do, I wouldn't have used the sticks until the day that I thought I was and then I would use them and I would be having a peak, so it became quite accurate.' (Interview with Nina)

Through observing various bodily sensations such as cramping, cervical fluids, libido, women came to feel and make sense of ovulation. They began to know when ovulation takes place and to tinker with the device to best test their bodies. Ovulation biosensors in this way came to occupy a secondary role (as a confirmer of ovulation), second to women's sensing of their bodies. They meant that women would select the best and most appropriate time to test – when they were sure they would obtain a positive result. Testing practices ultimately became more refined as women's bodies became the primary biosensor.

This finding diverges from historian Barbara Duden's (1992) notion of 'restricted sensorium' that suggests that women in contemporary societies come to know their bodies through medical professionals and accompanying technologies rather than through their own sensory experiences. An example of this is 'quickening': the moment when a woman feels the foetus move for the first time. Until the early 1900s, women's reports of quickening confirmed pregnancy. In the 20th century this was replaced by technologically mediated biomedical knowledge produced by pregnancy tests or ultrasound scans. Duden argues that such technologies restrict women's ability to sense their own bodies. In contrast, our findings show that ovulation biosensing (including online collective sense-making) can open up and expand the sensorium, in some cases helping women to sense ovulation for the first time.

Experimenting with ovulation biosensors

In addition to tracking ovulation, the women in Joann's study sometimes experimented with biosensors, using devices in ways in which they were not intended (Akrich, 1992; Oudshoorn, 2003; Wajcman, 2004). In so doing, they developed new understandings of oestrogen, progesterone and the luteinizing hormone (LH) and the ways in which these change their bodies over the ovulatory cycle. As women's knowledge grew deeper, they began to question their bodies and hormones, looking for new ways to test or try out their sensors. For example, some women learned that oestrogen changes and increases as pregnancy occurs. This led them to use the ovulation microscope to check for ferning patterns (denoting oestrogen) after ovulation which might indicate that a woman had conceived. Similarly, basal body temperature is typically higher during pregnancy; women would continue to take their temperatures after ovulation to check if they had become pregnant (temperatures remain high). In some cases, women tracked temperatures throughout pregnancy as a way of anticipating miscarriage, in which the basal body temperature would drop.

The ovulation microscope, and indeed microscopes in general, motivated women to experiment with bodily fluids and substances, for example, by looking at their male partner's saliva or, together with male partners, examining sperm under the microscope:

> I was curious about the saliva ferning as well. I actually work in a lab environment with microscopes so I thought it would be interesting to see if I could actually see the 'fern' pattern. I took a clean slide, spit on it (very ladylike, I know), let it dry, and then examined it under a microscope. While this isn't the best picture, here is what my saliva ferning looked like. I did this the day I ovulated (I knew I was ovulating by ovulation pain and a fertility monitor, also, my temp the following day confirmed it). Anyhow ... just thought I'd share. (Fertility forum user A)

> That's a great photo! Glad you shared ... now I'll know what I'm looking for. Maybe you'll know the answer to this question I asked on another forum. The [name of microscope] I have for Saliva Ferning

is I think 50x magnification. Any idea if that's strong enough to be able to see sperm? I saw some very large black circles with hollowed out centers ... but the more I think of it, they were way too big to be sperm, not to mention they didn't look at all like sperm and weren't moving. After looking at a few more slides online, I think what I saw was bubbles. Sperm even at 100x magnification would be tiny, right? (Fertility forum user B)

In this extract, User A determined that ovulation had taken place through a number of methods including pain, a fertility monitor and temperature tracking. She engaged with her laboratory microscope for fun and as a way to experiment. Microscopes and other ovulation biosensors allow for new ways to 'see' and know fertile bodies, which can be exciting in themselves, beyond their attachment to conception. Although FemTech are typically designed and marketed to track women's bodies, all bodies emerge as interesting and ultimately trackable in these experimental practices.

Ovulation biosensors are potentially feminist devices in that they also allow users to forge new connections between bodies and hormones beyond ovulation, fertility, trying to conceive or pregnancy. Hormones regulate bodies in so many different ways, yet their presence is felt more strongly in some narratives than in others, particularly around women's bodies and fertility. Like hormone trackers, ovulation biosensors may be of interest to those undergoing various periods of transition including perimenopause/ menopause and gender transition (Roberts and Waldby, 2021). In tracking hormones, users can push against the normative boundaries enacted by biosensing technologies to learn to see all kinds of bodies in different ways.

Conclusion

We began this chapter by asking how practices of ovulation biosensing build on earlier feminist experiments around women's health, bodies and the reshaping of science. We asked: how do such devices fit into feminist configurations of women's bodies and reproductive health, and bring new opportunities for health practices?

At the level of individual users, ovulation biosensors may be useful to some people for a period of time. They put ovulation on the map, locating this as an important process within bodies and fertilities, but also demonstrate that it is not 'the key' to conception that many manufacturers claim in their marketing materials (Wilkinson, 2016). Coming to know ovulation can be exciting, and part of a new 'journey' to becoming pregnant, and/or it can help to avoid unwanted pregnancies without using contraception (or limiting its use).

Ovulation biosensing also allows women to take part in the 'trying to conceive' worlds of online fertility forums. For many women, trying to conceive was a period of loneliness, isolation and uncertainty with few opportunities to work through these struggles with those close by. Online forums, in some ways, filled this gap, with some women participating in discussions on a daily basis. Women shared their personal histories of 'trying' and entered cycles of trying together, sharing and comparing ovulation data, symptoms and any oddities from that particular month. The online fertility forum combined with ovulation biosensors forums not only provided companionship but were also sources of information. If women were new to trying to conceive, they could post questions about which devices to use, how to use them and how to make sense of the data.

So, in what ways are ovulation biosensors feminist? In the 1960s and 1970s, widespread discrimination, misogyny and oppression mobilized women to come together to share experiences and learn about their bodies. There was an urgent political need to create new forms of knowledge and to do women's health differently, which ultimately led to women's health movements and organizations, such as the Boston Women's Health Collective and the Elizabeth Blackwell Health Center; the National Black Women's Health Project, SisterLove and Latina Roundtable. Although we argue that discriminatory and misogynistic practices around women's health persist, ovulation biosensing today typically has a different starting point: a desire to become pregnant (or to avoid becoming so) and to come to know ovulation in order to understand and manage fertility. Women begin sensing with the assumption that ovulation biosensors will tell them what they need to know. However, 'knowing ovulation' is not a simple, direct process, as bodies often change and ovulatory cycles may become irregular or affected by illness or stress. Additionally, data can be confusing, and hormones are not always trackable. Thus, knowledge of ovulation is not simply transmitted from device to user but instead is *done* through sensing practices such as the ones mentioned in this chapter. As in feminist self-help movements, women's practices of biosensing become a 'mobile set of practices, a mode for arranging knowledge production' (Murphy, 2012, p 25): women do science and create knowledge at home or in their private space with other women.

Ovulation biosensors offer potential as feminist technologies by making room for users to build expertise in self-sensing, to do (domestic) science and to become (collective) sense-makers. This is different from many fertility-tracking apps, which algorithmically process user-provided data to provide recommendations or 'solutions', largely bypassing women's bodies as sensors or sense-makers (and, indeed, women as knowledge-makers). Within fertility-tracking apps, women are configured as providers of information and objects of investigation rather than active participants in the creation of knowledge.

Manufacturers of fertility apps and ovulation biosensors might counter that their products make fertility tracking as easy as possible, thus minimizing potential misinterpretation of data. Equally, women may wish to know about ovulation without such practices of sense-making, which may ultimately become a source of burdensome reproductive labour. Approaching ovulation biosensing from a feminist STS perspective, as we have done here, however, shows clearly that when women use such devices, bodies continue to sense. Learning from this, manufacturers would do well, we suggest, to create technologies that work *with* and alongside bodies.

Furthermore, with practices of device-based ovulation biosensing, women's knowledge-building networks are opened up through interactions with other users on online forums in which women choose which data they share, when and with whom. There are significant differences as compared to fertility apps which tend to include online discussion forums as part of their 'offer'. These spaces are moderated by the companies who also provide interpretive guides and advice (see, for example, the webpages of 'Clue'). Fertility-tracking apps establish narrow, restrictive relationships between individual users and data companies. Data companies use women's data to expand their networks and profits through sharing with third parties (Tiffany, 2018). Arguably, they reduce women's sense-making worlds while expanding their own.

In spite of our assertion that ovulation biosensors offer potential for feminist health practices, we have serious concerns about such forms of capture and capitalization. The collective learning and sharing of experiences from the women's health movement was deeply embedded within the political agendas of feminism and improving women's (and others') lives (Murphy, 2012). As used in Joann's study, ovulation biosensors remained disconnected from political or social movements. Women come to (and leave) ovulation biosensing as individuals wanting, and in some cases struggling, to become pregnant. Their (in)fertilities are located in their own bodies, for which they are encouraged to take responsibility and seek individualized solutions. The social and political agendas surrounding caring for reproductive bodies and lives in contemporary society are largely absent here: to date, ovulation-biosensing communities have not produced collective demands for better care or support for women's reproductive lives. The rising public criticisms of the corporatization of menstrual-tracking apps and increasing experimentation with hormonal biosensing may, however, hold the seeds of such demands. We hope feminist STS research in this space will facilitate this change.

Notes

[1] An example of the device can be seen at www.fairhavenhealth.com/fertile-focus

[2] An example can be seen at https://homehealth-uk.com/all-products/one-step-10-x-3-5mm-ovulation-test-strips-20miuml/

[3] An example can be seen at www.zoombaby.co.uk/product/basal-thermometer-and-bbt-charting-pack/

References

Akrich, M. (1992) 'The de-scription of technical objects', in W.E. Bijker and J. Law (eds) *Shaping Technology/Building Society*, Cambridge, MA: MIT Press.

Balasch, J. and Gratacós, E. (2012) 'Delayed childbearing: effects on fertility and the outcome of pregnancy', *Current Opinion in Obstetrics and Gynecology* 24(3): 187–193.

Bojar, K. (2021) *The Women's Health Movement*, New York: Routledge

Boston Women's Health Book Collective (2011) *Our Bodies, Ourselves*, New York: Touchstone, Simon & Schuster.

Campbell, D. (2014) 'Women who postpone having babies may stay childless, warns medical chief', *The Guardian,* www.theguardian.com/society/2014/jan/17/women-postpone-babies-childless-medical-chief.

Chen, S.H. and Wallach, E.E. (1994) 'Five decades of progress in management of the infertile couple', *Fertility and Sterility*, 62(4): 665–685.

Crawford, K., Lingel, J. and Karppi, T. (2015) 'Our metrics, ourselves: a hundred years of self-tracking from the weight scale to the wrist wearable device', *European Journal of Cultural Studies*, 18(4–5): 479–496.

Davies, W. (2015) *The Happiness Industry: How the Government and Big Business Sold Us Well-being*, London: Verso.

Duden, B. (1992) 'Quick with child: an experience that has lost its status', *Technology in Society*, 14(3): 335–344.

Faddy, M.J. and Godsen, R.G. (2003) 'Modelling the dynamics of ovarian follicle utilization throughout life', in A.O. Trounson and R.G. Godsen (eds) *Biology and Pathology of the Oocyte: Role in Fertility and Reproductive Medicine*, Cambridge: Cambridge University Press, pp 44–52.

Faulkner, A. (2009) *Medical Technology into Healthcare and Society: A Sociology of Devices, Innovation and Governance*, Basingstoke: Palgrave Macmillan.

Federal Trade Commission (2021) 'Developer of popular women's fertility-tracking app settles FTC allegations that it misled consumers about the disclosure of their health data', www.ftc.gov/news-events/press-releases/2021/01/developer-popular-womens-fertility-tracking-app-settles-ftc.

Federation of Feminist Women's Health Centers (1981) *How to Stay Out of the Gynecologist's Office*, Los Angeles, CA: Women to Women Publications.

Franklin, S. (1997) *Embodied Progress: A Cultural Account of Assisted Conception*, London and New York: Routledge.

Franklin, S. and Roberts, C. (2006) *Born and Made: An Ethnography of Preimplantation Genetic Diagnosis*, Oxford: Princeton University Press.

Gambier-Ross, K., McLernon, D.J. and Morgan, H.M. (2018) 'A mixed methods exploratory study of women's relationships with and uses of fertility tracking apps', *Digital Health*, 4: 1–15.

Glow (2012) Revolutionizing women's health through data & AI, https://glowing.com/about-glow/

Grenfell, P. et al (2021) 'Fertility and digital technology: NARRATIVES of using smartphone app "Natural Cycles" while trying to conceive', *Sociology of Health & Illness*, 43(1): 116–132.

Grundy, Q., Chiu, K., Held, F., Continella, A., Bero, L. and Holz, R. (2019) 'Data sharing practices of medicines related apps and the mobile ecosystem: traffic, content, and network analysis', *BMJ*, 364: 1920.

Hamper, J. (2020) ' "Catching ovulation": exploring women's use of fertility tracking apps as a reproductive technology', *Body & Society* 26(3): 3–30.

Haraway, D.J. (1997) *Modest Witness @Second Millennium: FemaleMan Meets OncoMouse: Feminism and Technoscience*, New York and London: Routledge.

Harding, S.G. (1986) *The Science Question in Feminism*, Milton Keynes: Open University Press.

Harris, A., Kelly, S. and Wyatt, S. (2016) *CyberGenetics: Health Genetics and New Media*, London: Routledge.

Healy, R.L. (2021) 'Zuckerberg, get out of my uterus! An examination of fertility apps, data-sharing and remaking the female body as a digitalized reproductive subject', *Journal of Gender Studies*, 30(4): 406–416.

Hendl, T., Jansky, B. and Wild, V. (2019) 'From design to data handling: why mHealth needs a feminist perspective', in J. Loh and M. Coeckelbergh (eds) *Feminist Philosophy of Technology*, Stuttgart: J.B. Metzler, pp 77–103.

Hope, J. (2013) 'Dangers of delaying motherhood until 30: don't think the risks begin at 35, say researchers', www.dailymail.co.uk/health/article-2521 666/Dangers-delaying-motherhood-30.html.

Johnson, S. (2014) 'Maternal devices: social media and the self-management of pregnancy, mothering and child health', *Societies*, 4(2): 330–350.

Kenner, A.M. (2008) 'Securing the elderly body: dementia, surveillance and the politics of "aging in place"', *Surveillance and Society*, 5(3): 252–69.

Kleinman, Z. (2021) 'Are women being let down by period trackers?' BBC News, 17 January, https://www.bbc.com/news/technology-55146149.

Kline, W. (2010) *Bodies of Knowledge: Sexuality, Reproduction, and Women's Health in the Second Wave*, Chicago: University of Chicago Press.

Kragh-Furbo, M. (2015) 'Doing susceptibilities: genetic data practices in consumer genomics', PhD thesis, Lancaster University, Lancaster, UK.

Kragh-Furbo, M. and Tutton, R. (2017) 'Spitting images: remaking saliva as a promissory substance', *New Genetics and Society*, 36(2): 159–185.

Kressbach, M. (2021) 'Period hacks: menstruating in the big data paradigm', *Television & New Media*, 22(3): 241–261.

Layne, L. (2009) 'The home pregnancy test: A feminist technology?' *Women's Studies Quarterly*, 37(1/2): 61–79.

Leavitt, A.S. (2006) 'A private little revolution: the home pregnancy test in American culture', *Bulletin of the History of Medicine*, 80(2): 317–345.

Ledger, D. and McCaffrey, D. (2014) 'Inside wearables: how the science of human behavior change offers the secret to long-term engagement', *Endeavour Partners*, 200(1): 1–17.

Lupton, D. (2013) 'Quantifying the body: monitoring and measuring health in the age of mHealth technologies', *Critical Public Health*, 23(4): 393–403.

Lupton, D. (2014a) 'The commodification of patient opinion: the digital patient experience economy in the age of big data', *Sociology of Health & Illness*, 36(6): 856–869.

Lupton, D. (2014b) 'Quantified sex: a critical analysis of sexual and reproductive self-tracking using apps', *Culture, Health & Sexuality*, 17(4): 1–14.

Mamo, L. (2007) *Queering Reproduction: Achieving pregnancy in the Age of Technoscience*, Durham, NC: Duke University Press.

Martin, L.J. (2010) 'Anticipating infertility: egg freezing, genetic preservation, and risk', *Gender & Society*, 24(4): 526–545.

Mort, M., Roberts, C., Kragh-Furbo, M., Wilkinson, J. and Mackenzie, A. (2016) 'Biosensing: how citizens' views illuminate emerging health and social risks', *Health, Risk & Society*, 17(7–8): 605–23.

Murphy, M. (2004) 'Immodest witnessing: the epistemology of vaginal self-examination in the US feminist self-help movement', *Feminist Studies*, 30(1): 115–147.

Murphy, M. (2012) *Seizing the Means of Reproduction: Entanglements of Feminism, Health, and Technoscience*, Durham, NC: Duke University Press.

Nafus, D. (2016) *Quantified: Biosensing Technologies in Everyday Life*, Cambridge, MA: MIT Press.

Nafus, D. and Sherman, J. (2014) 'The quantified self movement as an alternative big data practice', *International Journal of Communication*, 8: 1785–1794.

Nahman, M. (2011) 'Reverse traffic: intersecting inequalities in human egg donation', *Reproductive Biomedicine Online*, 23(5): 626–633.

Neff, G. and Nafus, D. (2016) *Self-tracking*, Cambridge, MA: MIT Press.

Nelson, C. (1977) *Self-help Home Remedies*, Chico, CA: Chico Feminist Women's Health Center.

Nordqvist, P. (2011) 'Dealing with sperm: comparing lesbians' clinical and non-clinical donor conception processes', *Sociology of Health and Illness*, 33(1): 114–129.

Oudshoorn, N. (2003) *The Male Pill: A Biography of a Technology in the Making*, Durham, NC: Duke University Press.

Pantzar, M. and Ruckenstein, M. (2015) 'The heart of everyday analytics: emotional, material and practical extensions in self-tracking market', *Consumption, Markets and Culture*, 18(1): 92–109.

Ragone, H. (1994) *Surrogate Motherhood: Conception in the Heart*, Boulder, CO: Westview Press.

Rapp, R. (1999) *Testing Women, Testing the Fetus: The Social Impact of Amniocentesis in America*, New York: Routledge.

Rettberg, J.W. (2014) *Seeing Ourselves through Technology: How We Use Selfies, Blogs and Wearable Devices to See and Shape Ourselves*, London: Palgrave Macmillan.

Roberts, C. and Waldby, C. (2021) 'Incipient infertility: tracking eggs and ovulation across the life course', *Catalyst,* 7(1): 1–25.

Roberts, C., Mackenzie, A., Mort, M., with Atkinson, T., Kragh-Furbo, M. and Wilkinson, J. (2019) *Living Data: Making Sense of Health Biosensing*, Bristol: Bristol University Press.

Russell, H. (2019) 'In search of "smart chewing gum" to help women conceive', *The Guardian*, 10 October. https://www.theguardian.com/society/2019/oct/10/in-search-of-smart-chewing-gum-to-help-women-conceive.

Sanders, R. (2017) 'Self-tracking in the digital era', *Body & Society*, 23(1): 36–63.

Savage, M. (2017) 'The Swedish physicist revolutionising birth control', BBC News, www.bbc.co.uk/news/business-40629994.

Shadwell, T. (2019) 'Period tracker app spied on me – and told advertisers it thought I was pregnant' *Daily Mirror*, 4 Nov, https://www.mirror.co.uk/news/uk-news/period-tracker-app-spied-told-20807187.

Smith, D.E. (1990) *The Conceptual Practices of Power: A Feminist Sociology of Knowledge*, Boston, MA: Northeastern University Press.

Takhar, J. and Pemberton, K. (2019) 'Reproducing "rhetrickery" in online fertility marketing: harnessing the "rhetoric of the possible"', *Consumption, Markets & Culture*, 22(4): 314–336.

Thomas, S.L., Nafus, D. and Sherman, J. (2018) 'Algorithms as fetish: faith and possibility in algorithmic work', *Big Data & Society*, 5(1): 1–11.

Thompson, C. (2005) *Making Parents: The Ontological Choreography of Reproductive Technologies*, Cambridge, MA: MIT Press.

Throsby, K. (2004) *When IVF Fails: Feminism, Infertility and the Negotiation of Normality*, Basingstoke: Palgrave Macmillan.

Tiffany, K. (2018) 'Period-tracking apps are not for women', *Vox*, www.vox.com/the-goods/2018/11/13/18079458/menstrual-tracking-surveillance-glow-clue-apple-health.

Wajcman, J. (2004) *Technofeminism*, Cambridge: Polity Press.

Waldby, C. (2015) 'Banking time: egg freezing and the negotiation of future fertility', *Culture, Health and Sexuality*, 17(4): 470–482.

Weigel, M. (2016) '"Fitbit for your period": the rise of fertility tracking', *The Guardian*, www.theguardian.com/technology/2016/mar/23/fitbit-for-your-period-the-rise-of-fertility-tracking.

Wilkinson, J. (2016) 'Ovulation biosensing: emerging practices of sensing and knowing', PhD thesis, Lancaster University, Lancaster, UK.

Wilkinson, J. (2020) 'Technologies of time: women's practices of trying to conceive with ovulation biosensing', *Sociology of Health & Illness*, 42(7): 1597–1610.

Wilkinson, J., Roberts, C. and Mort, M. (2015) 'Ovulation monitoring and reproductive heterosex: living the conceptive imperative?' *Culture, Health & Sexuality*, 17(4): 454–469.

Conclusion

What is new, is how social and behavioural genomics research is converging with systems of education to produce sociotechnic imaginaries that either enhance or further diminish the sociopolitical visibility of marginalized groups. (Daphne Oluwaseun Martschenko, Chapter 1)

[T]his chapter has attended to and explored important sites of disjuncture, inconsistency and potential rupture around what it means to be in 'good health'. This, I argue, can function as sources of opportunity for rethinking health and challenging the hidden dimensions of inequality that continue to be reproduced. (Tina Sikka, Chapter 2)

[W]hile research on the biomedicalization of HIV has generated a rich understanding of the complexities of living with HIV, the 'post crisis' framing (at a time of global austerity), occludes the significant challenges that continue to be faced by HIV-positive people, and by HIV-positive women in particular. (Elizabeth Mills, Chapter 3)

[O]ne of the answers to the question of why the potential threats and fears of artefacts do not disappear seems to be largely due to the fact that the development of such technologies prioritizes profits, neglects the risk assessment from the perspective of the consumers and does not establish disabling technologies against the target technology. (Kazuhiko Shibuya, Chapter 4)

AI could be a deliberative and caregiving partner in future pandemics that might help individual patients avoid the isolation, disconnection and fear that were experienced so significantly over the past year and a half by so many while creating safer clinical spaces. (Kevin Cummings and John Rief, Chapter 5)

Researchers and practitioners in science communication need to prioritize inclusion, equity and intersectionality in order to engage and benefit the entire community with science communication. (Anamika Gulati, Chapter 6)

Far from the idea that associating negative concepts of authority and control with technology can actually warn against the adverse effects of technological changes, which could also lead to positive effects ... (Filomena Berardi and Giorgio Vernoni, Chapter 7)

If robots are to be developed, it is essential to involve in the process the different actors who, in one way or another, are involved in their use (patients, relatives, caregivers, health professionals and so on) and to find ways to enable their collaboration with engineers and designers. (Miquel Domènech and Núria Vallès-Peris, Chapter 8)

The social and political agendas surrounding caring for reproductive bodies and lives in contemporary society are largely absent here: to date, ovulation-biosensing communities have not produced collective demands for better care or support for women's reproductive lives. (Joann Wilkinson and Celia Roberts, Chapter 9)

The quotes listed here have been taken directly from the tail end of each of the chapters that comprise this book. In addition to summarizing their central arguments, these quotes are ones that I happened to have starred or underlined for capturing the essence of contemporary STS scholarship – specifically its turn towards analyses tied to the eradication of marginalization, oppression and inequality in favour of a world that is intersectional, collective and just. I also felt that gathering them together in this way would do the work of 'summarization' while also allowing me to perform one last agential cut of my own into this contingently assembled assemblage of chapters. Remember that an agential cut, according to Karen Barad, involves the temporary stabilization of messy entanglements so as to render 'it' knowable but not exploitable. Cutting of this sort 'is [thus] not about othering or separating but on the contrary about making connections and commitments' (Barad, 2007, p 392). Each of these chapters performs and enacts a 'commitment' of this kind – particularly as it relates to the cultivation of just outcomes propelled by the variegated STS methods, theories, histories and knowledges that have been deployed. I would add that this understanding of STS also reflects a shared, careful assessment of how our discourses, materialities and embodiments intra-act with agential non-human technologies (broadly defined) either to co-produce or to undermine just outcomes.

Also uniting these chapters is how STS has been used to challenge and critique studies that, in the past, would have passed without notice. Martschenko's analysis in Chapter 1 of the problematic assumptions of contemporary genomics research in education, Gulati's assessment in Chapter 6 of the blockages in contemporary science communication, and my own Chapter 2 challenging the neoliberal conception of health deploy this approach concretely. They do this work by judiciously engaging with knowledge and practices that can be harmful to marginalized groups – ones that 'oversimplify and misuse the categories of race, sex, gender, class, and more' (Fitsch et al, 2020, p 4). Justice and equity form the ethico-political frameworks of this collection wherein each chapter, in its own way, identifies and struggles against techno-social relations that are appropriative and alienating, in favour of ones that highlight solidarity, Otherness and good relations.

In addition to a turn towards care and justice, these quotations also reveal that not everyone in the collection is on the same page vis-à-vis our technical futures. As noted in the Introduction, the positionalities of each contributor span more positive iterations of technological futures, such as the contribution of Berardi and Vernoni (Chapter 7) and Domènech and Vallès-Peris (Chapter 8), to more critical assessments from, for example, Martschenko (Chapter 1) and myself (Chapter 2). I would argue, however, that the oft-used utopian/dystopian dichotomy is unhelpful here since each chapter is significantly more nuanced and aims at unpacking how, for example, future health technologies might potentially cultivate medical relationality, as Cummings and Rief argue (Chapter 5).

Finally, the collected 'sensibilities' of these chapters offer important ways for STS to continue on this path. These sensibilities include: (1) that health and wellness are socially constructed; (2) that technologies are agential (as in the case of antiretrovirals and biosensors); and (3) that the semiotic materiality of self, other and technology are co-constituted (see Wajcman, 2006; Pinch, 2009; Henwood and Marent, 2019). This kind of practice ensures that space is made for what Sismondo describes as the most important contribution of STS, namely, to 'create space for theoretically sophisticated analyses of science and technology in explicitly political contexts' (Sismondo, 2018, p 356).

In closing, I would encourage readers to continue to follow the work of each of these scholars whose reflexive and discerning insights reflect what is most exciting about this collection.

References

Barad, K. (2007) *Meeting the Universe Halfway*, Durham, NC: Duke University Press.

Fitsch, H., Jordan-Young, R., Trujillo, A.K., Kraus, C., Roy, D. and Schmitz, S. (2020) 'Coalition-making and the practice of feminist STS in the time of COVID-19', *Catalyst: Feminism, Theory, Technoscience*, 6(2): 1–32.

Henwood, F. and Marent, B. (2019) 'Understanding digital health: productive tensions at the intersection of sociology of health and science and technology studies', *Sociology of Health & Illness,* 41: 1–15.

Pinch, T. (2009) 'The social construction of technology (SCOT): the old, the new, and the nonhuman', in P. Vannini (ed) *Material Culture and Technology in Everyday Life: Ethnographic Approaches*, New York: Peter Lang, pp 45–58.

Sismondo, S. (2018) 'Science and technology studies', in N. Castree, M. Hulme and J.D. Proctor (eds) *Companion to Environmental Studies*, London: Routledge, pp 356–359.

Wajcman, J. (2006) 'Technocapitalism meets technofeminism: women and technology in a wireless world', *Labour & Industry: A Journal of the Social and Economic Relations of Work*, 16(3): 7–20.

Index

References to endnotes show both the page number
and the note number (231n3).